Simula SpringerBriefs on Computing

Volume 15

In 2016, Springer and Simula launched the book series *Simula SpringerBriefs on Computing*, which aims to provide introductions to selected research topics in computing. The series provides compact introductions for students and researchers entering a new field, brief disciplinary overviews of the state-of-the-art of select fields, and raises essential critical questions and open challenges in the field of computing. Published by SpringerOpen, all *Simula SpringerBriefs on Computing* are open access, allowing for faster sharing and wider dissemination of knowledge.

Simula Research Laboratory is a leading Norwegian research organization which specializes in computing. Going forward, the book series will provide introductory volumes on the main topics within Simula's expertise, including communications technology, software engineering and scientific computing.

By publishing the *Simula SpringerBriefs on Computing,* Simula Research Laboratory acts on its mandate of emphasizing research education. Books in this series are published by invitation from one of the series editors. Authors interested in publishing in the series are encouraged to contact any member of the editorial board.

Joakim Sundnes

Solving Ordinary Differential Equations in Python

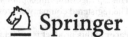

Joakim Sundnes
Simula Research Laboratory AS
Oslo, Norway

ISSN 2512-1677　　　　　　　　ISSN 2512-1685　(electronic)
Simula SpringerBriefs on Computing
ISBN 978-3-031-46767-7　　　　ISBN 978-3-031-46768-4　(eBook)
https://doi.org/10.1007/978-3-031-46768-4

Mathematics Subject Classification: 34-01, 34-04, 65-L05, 65-L04, 65-L06, 39-08

This Springer imprint is published by the registered company Springer Nature Switzerland AG
The registered company address is: Gewerbestrasse 11, 6330 Cham, Switzerland

Paper in this product is recyclable.

Series Foreword

Dear reader,

Scientific research is increasingly interdisciplinary, and both students and experienced researchers often face the need to learn the foundations, tools, and methods of a new research field. This process can be quite demanding, and typically involves extensive literature searches and reading dozens of scientific papers in which the notation and style of presentation varies considerably. Since the establishment of this series in 2016 by founding editor-in-chief Aslak Tveito, the briefs in this series have aimed to ease the process by introducing and explaining important concepts and theories in a relatively narrow field, and to outline open research challenges and pose critical questions on the fundamentals of that field. The goal is to provide the necessary understanding and background knowledge and to motivate further studies of the relevant scientific literature. A typical brief in this series should be around 100 pages and should be well suited as material for a research seminar in a well-defined and limited area of computing.

We publish all items in this series under the SpringerOpen framework, as this allows authors to use the series to publish an initial version of their manuscript that could subsequently evolve into a full-scale book on a broader theme. Since the briefs are freely available online, the authors do not receive any direct income from the sales; however, remuneration is provided for every completed manuscript. Briefs are written on the basis of an invitation from a member of the editorial board. Suggestions for possible topics are most welcome and can be sent to sundnes@simula.no.

March 2023

Dr. Joakim Sundnes
Editor-in-Chief
Simula Research Laboratory
sundnes@simula.no

Dr. Martin Peters
Executive Editor Mathematics
Springer Heidelberg, Germany
martin.peters@springer.com

Series editor for this volume

Aslak Tveito, Simula Research Laboratory, Oslo, Norway

Preface

This book was based on a set of lecture notes originally written for the book *A Primer on Scientific Programming with Python* by Hans Petter Langtangen [14], mainly covering topics from Appendices A, C, and E. To provide a more comprehensive overview of state-of-the art solvers for ordinary differential equations (ODEs), the notes have been extended with additional material on implicit solvers and automatic time-stepping methods. The main purpose of the notes is to serve as a concise and gentle introduction to solving differential equations in Python, specifically for the course *Introduction to programming for scientific applications* (IN1900, 10 ETCS credits) at the University of Oslo. These notes will be most useful for readers with a basic knowledge of Python and NumPy, see for instance [16], and it is also useful to have a fundamental understanding of ODEs.

One may question the usefulness of learning how to write your own ODE solvers in Python when there are already multiple solvers available, such as those in the SciPy library. However, no single ODE solver is universally optimal and efficient for all ODE problems, and the choice of solver should always be based on the specific characteristics of the problem at hand. To make the right choice, it is extremely beneficial to understand the strengths and weaknesses of different solvers, and the best way to gain this knowledge is by programming your own collection of ODE solvers. Different ODE solvers are conveniently grouped into families and hierarchies, offering an excellent example of how object-oriented programming (OOP) can maximize code reuse and minimize duplication.

The book's presentation style is compact and pragmatic, incorporating numerous code examples to illustrate how various ODE solvers can be implemented and applied in practice. The complete source code for all examples, as well as Jupyter notebooks for each chapter, are provided in the accompanying online resources. The programs and code examples are written in a simple and compact Python style, avoiding the use of advanced tools and features. Experienced Python programmers may find more elegant and modern solutions to many of the examples, utilizing abstract base classes, type hints, data classes, and other advanced features. However, the book's main goal is to introduce the fundamentals of ODE solvers and OOP as part of an introductory programming course, and we believe this purpose is best served by focusing on the basics.

Readers familiar with scientific computing or numerical software may also miss a discussion of computational performance. While performance is certainly relevant when solving ODEs, optimizing the performance of a Python-based solver easily becomes quite technical, and requires features like just-in-time compilers (e.g., Numba) or mixed-language programming. The solvers in this book use fairly basic features of Python and NumPy, sacrificing some performance in favor of enhancing understanding of solver properties and implementation.[1]

The book is organized as follows: Chapter 1 introduces the forward Euler method, serving as a foundation for understanding the principles underlying all the methods covered later. It introduces the notation and mathematical formulation used throughout the book for scalar ODEs and systems of ODEs, and is essential reading for those with limited prior experience with ODEs and ODE solvers. Additionally, it briefly explains how to use the ODE solvers from the SciPy library. Readers already familiar with the fundamentals of the forward Euler method and its implementation may consider proceeding straight to Chapter 2, which presents explicit Runge-Kutta methods. The chapter introduces the fundamental ideas of these methods, but the main focus is on the implementation and how a collection of ODE solvers is conveniently implemented as a class hierarchy. Chapter 3 introduces *stiff* ODEs, presents techniques for performing simple stability analysis of Runge-Kutta methods, and introduces implicit Runge-Kutta methods. The majority of the chapter is dedicated to the programming of these solvers, which exhibit better stability properties than explicit methods and are therefore more suitable for solving stiff ODEs. Chapter 4 concludes the presentation of ODE solvers by introducing methods for adaptive time step control, which is an essential component of all modern ODE software. Chapter 5 takes a different approach from the preceding chapters, as it focuses on a specific class of ODE models rather than a set of solvers. While the simpler ODE problems discussed in earlier chapters serve the purpose of introducing and testing the solvers, it is valuable to explore more complex ODE models in order to appreciate both the potential and the challenges of modeling with ODEs. As an example, the chapter examines the famous Kermack-McKendrick SIR (Susceptible-Infected-Recovered) model from epidemiology. These classic models were developed in the early 1900s (see [12]) and remain fundamental for predicting and understanding the spread of infectious diseases. We describe the derivation of the models from a set of fundamental assumptions, and discuss the implications and limitations resulting from these assumptions. The main focus of the chapter is then on modifying and extending the models to capture new phenomena, and

---

[1]Complete source code for all the solvers and examples in the book can be found here: https://sundnes.github.io/solving_odes_in_python/

demonstrating how these changes can be implemented and explored using the solvers developed in preceding chapters.

Finally, while the main focus of the text is on differential equations, Appendix A is dedicated to the related topic of *difference equations*. Difference equations have important applications on their own and may serve as a stepping stone towards understanding and solving ODEs, since numerical methods for ODEs essentially involve transforming differential equations into difference equations. The standard formulation of difference equations found in mathematical textbooks is already well-suited for computer implementation, using for-loops and arrays. Some students find difference equations easier to grasp than differential equations, making Appendix A a useful resource to begin with. However, others may prefer to dive straight into ODEs and explore Appendix A at a later stage.

July 2023   *Joakim Sundnes*

Contents

Chapter 1
Programming a Simple ODE Solver

Ordinary differential equations (ODEs) are widely used in science and engineering, particularly when it comes to modeling dynamic processes. Although analytical methods can be employed to solve simple ODEs, nonlinear ODEs typically require numerical methods for solutions. In this chapter we demonstrate how to program general numerical solvers capable of handling any ODE. Initially we will focus on scalar ODEs, which consist of a single equation and a single unknown. Subsequently, in Section 1.3, we will extend these concepts to systems of coupled ODEs. Acquiring a solid grasp of the concepts presented in this chapter will not only help you with programming your own ODE solvers but also in using a diverse range of readily available, general-purpose ODE solvers in Python or other programming languages.

1.1 Creating a General-Purpose ODE Solver

When solving ODEs analytically, one typically considers a specific ODE or a class of ODEs and attempts to derive a formula for the solution. However, in this chapter, our goal is to implement numerical solvers that can be applied to any ODE, without being limited to a single example or a specific class of equations. To achieve this, we need a general abstract notation for an arbitrary ODE. We will write the ODEs on the following form:

$$u'(t) = f(t, u(t)), \tag{1.1}$$

which means that the ODE is fully specified by the definition of the right-hand side function $f(t, u)$. Examples of this function may be:

© The Author(s) 2024
J. Sundnes, *Solving Ordinary Differential Equations in Python*,
Simula SpringerBriefs on Computing 15,
https://doi.org/10.1007/978-3-031-46768-4_1

$$f(t, u) = \alpha u, \quad \text{exponential growth}$$

$$f(t, u) = \alpha u \left(1 - \frac{u}{R}\right), \quad \text{logistic growth}$$

$$f(t, u) = -b|u|u + g, \quad \text{falling body in a fluid}$$

Notice that, for the sake of generality, we write all the right-hand sides of the ODEs as functions of both t and u, even though the mathematical formulations only involve u. This general formulation is not strictly necessary in the mathematical equations, but it proves to be highly convenient when we start programming and want to use the same solver for a diverse range of ODE models. We will delve into this topic in greater detail later. Now, our objective is to write functions and classes that accept the function f as input and solve the corresponding ODE to generate the output u.

To ensure a unique solution for (1.1), it is necessary to specify the *initial condition* for u. This initial condition corresponds to the value of the solution at a specific time $t = t_0$. The resulting mathematical problem can be expressed as

$$u'(t) = f(t, u(t)),$$
$$u(t_0) = u_0,$$

and is commonly referred to as an *initial value problem*, or simply an IVP. Every ODE problem discussed in this book is an initial value problem. To illustrate, let us consider the very simple ODE

$$u' = u.$$

This general solution of this equation is given by $u(t) = Ce^t$ for any constant C, implying that there exist an infinite number of solutions. However, by specifying an initial condition $u(t_0) = u_0$, we get $C = u_0$ and the unique solution $u(t) = u_0 e^t$. When solving the equation numerically, it is necessary to define the initial condition u_0 in order to start our method and compute a solution at all.

A Simple and General Solver: the Forward Euler Method. A numerical method for (1.1) can be derived by using a finite difference approximation for the derivative in the equation $u' = f(t, u)$. To introduce this idea, let us assume that we have already computed u at discrete time points $t_0, t_1, \ldots, t_n$. At time t_n we have the ODE

$$u'(t_n) = f(t_n, u(t_n)),$$

and we can now approximate $u'(t_n)$ with a forward finite difference:

$$u'(t_n) \approx \frac{u(t_{n+1}) - u(t_n)}{\Delta t}.$$

By inserting this approximation into the ODE at $t = t_n$, we obtain the following equation

$$\frac{u(t_{n+1}) - u(t_n)}{\Delta t} = f(t_n, u(t_n)),$$

and we can rearrange the terms to obtain an explicit formula for $u(t_{n+1})$:

$$u(t_{n+1}) = u(t_n) + \Delta t f(t_n, u(t_n)).$$

This method, known as the *Forward Euler (FE) method* or the *Explicit Euler method*, is the simplest numerical method for solving an ODE. The terms *forward* and *explicit* refer to the fact that we have an explicit update formula for $u(t_{n+1})$ that only involves known quantities at time t_n. In contrast, an *implicit* ODE solver would have an update formula that includes terms like $f(t_{n+1}, u(t_{n+1}))$, requiring the solution of a generally nonlinear equation to determine the unknown $u(t_{n+1})$. We will explore other explicit ODE solvers in Chapter 2 and implicit solvers in Chapter 3.

To simplify the formula, we introduce the notation $u_n = u(t_n)$, i.e., we let u_n represent the numerical approximation to the exact solution $u(t)$ at $t = t_n$. With this notation, the update formula reads

$$u_{n+1} = u_n + \Delta t f(t_n, u_n), \tag{1.2}$$

which, if we know the u_0 at time t_0, can be applied repeatedly to u_1, u_2, u_3 and so forth. If we again consider the very simple ODE given by $u' = u$ (i.e., $f(t, u) = u$), we have

$$u_1 = u_0 + \Delta t u_0,$$
$$u_2 = u_1 + \Delta t u_1,$$
$$u_3 = u_2 + \ldots,$$

and the general update formula

$$u_{n+1} = u_n + \Delta t u_n = (1 + \Delta t) u_n.$$

In a Python program, the repeated application of the same formula can be conveniently implemented using a for-loop, and the solution can be stored in a list or a NumPy array. If you are unfamiliar with NumPy arrays and their usage, we recommend referring to [16], which provides an introduction to NumPy arrays and tools that will be extensively used through this book. To solve the ODE numerically, given a final time T and the number of time steps N, we can follow these steps:

1. Create arrays t and u of length $N + 1$[1]
2. Set initial condition: u[0] $= u_0$, t[0] $= 0$

---

[1] For N time steps, the length of the arrays needs to be $N + 1$ since we need to store both end points, i.e., $t_0, t_1, \ldots, t_n$ and $u_0, u_1, \ldots, u_n$.

3. Compute the time step Δt dt = T/N
4. For $n = 0, 1, 2, \ldots, N-1$:

- t[n + 1] = t[n] + dt
- u[n + 1] = (1 + dt) * u[n]

A complete Python implementation of this algorithm may look like

```
import numpy as np
import matplotlib.pyplot as plt

N = 20
T = 4
dt = T/N
u0 = 1

t = np.zeros(N + 1)
u = np.zeros(N + 1)

u[0] = u0
for n in range(N):
    t[n + 1] = t[n] + dt
    u[n + 1] = (1 + dt) * u[n]

plt.plot(t, u)
plt.show()
```

Notice that there is no need to set t[0]= 0 when t is created in this way, but it is important to update u[0]. Forgetting to do so is a common error in ODE programming, so it is worth taking note of the line u[0] = u0. The solution is shown in Figure 1.1 for two different choices of the time step Δt. As observed, the approximate solution improves as Δt is reduced, although both the solutions deviate from the exact solution. However, reducing the time step further would easily yield a solution that is indistinguishable from the exact solution.

The for-loop in the aforementioned example could also be implemented differently, for instance

```
for n in range(1, N+1):
    t[n] = t[n - 1] + dt
    u[n] = (1 + dt) * u[n - 1]
```

Here, the index n runs from 1 to N, and all the indices inside the loop have been decreased by one to achieve the same outcome. In this simple case, it is easy to verify that both loop formulations give the same result. However, mixing up the two formulations can easily lead to errors, such as a loop that exceeds the array bounds (resulting in an IndexError) or a loop where the last elements of t and u are not computed. Although these errors may appear trivial, they are common pitfalls when working with for-loops and it is good practice to always examine the loop formulation to ensure consistent use of indices and bounds.

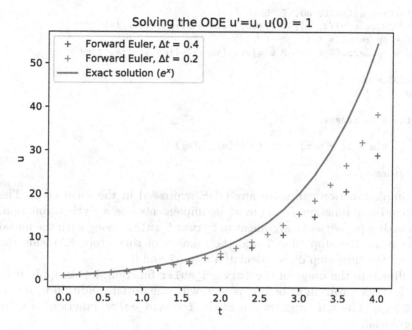

Fig. 1.1 Solution of $u' - u, u(0) = 1$ with $\Delta t = 0.4$ ($N = 10$) and $\Delta t - 0.2$ ($N = 20$).

Extending the Solver to the General ODE. As mentioned earlier, the goal of this chapter is to develop general-purpose ODE solvers capable of solving any ODE expressed in the form $u' = f(t, u)$. Achieving this requires only a slight modification of the algorithm presented above:

1. Create arrays t and u of length $N+1$
2. Set initial condition: u[0] $= u_0$, t[0]=0
3. For $n = 0, 1, 2, \ldots, N-1$:

   - t[n + 1] = t[n] + dt
   - u[n + 1] = u[n] + dt * f(t[n], u[n])

The modified version of the algorithm only requires a small change in the formula for computing u[n+1] from u[n]. In the previous case we had $f(t, u) = u$, and to create a general-purpose ODE solver we simply replace u[n] with the more general f(t[n],u[n]). The following Python function implements this generic version of the FE method:[2]

```python
import numpy as np

def forward_euler(f, u0, T, N):
    """Solve u'=f(t, u), u(0)=u0, with n steps until t=T."""
    t = np.zeros(N + 1)
    u = np.zeros(N + 1)  # u[n] is the solution at time t[n]

    u[0] = u0
    dt = T / N

    for n in range(N):
        t[n + 1] = t[n] + dt
        u[n + 1] = u[n] + dt * f(t[n], u[n])

    return t, u
```

This simple function can solve any ODE expressed in the form (1.1). The right-hand side function $f(t, u)$ must be implemented as a Python function, which is then passed as an argument to forward_euler, along with the initial condition u0, the stop time T and the number of time steps N. Inside the function, the time step dt is calculated using T and N.

To illustrate the usage of the forward_euler function, let us apply it to solve the same problem as before: $u' = u$, with the initial condition $u(0) = 1$, for $t \in [0, 4]$. The following code uses the forward_euler function to solve this problem:

```python
def f(t, u):
    return u

u0 = 1
T = 4
N = 30
t, u = forward_euler(f, u0, T, N)
```

The forward_euler function returns two arrays, t and u, which can be further processed or plotted as desired. An important aspect to note in this code is the definition of the right-hand side function f. As mentioned earlier, this function should always be written with two arguments, t and u, although in this case only u is used inside the function. The inclusion of both arguments is necessary because we want our solver to be applicable for all ODEs in the form $u' = f(t, u)$. Therefore, inside the forward_euler function, the f function is called as f(t[n], u[n]). If the right-hand side function were defined as a function of u only, i.e., using def f(u):, an error would occur when

---

[2]The source code for this function, as well as all subsequent solvers and examples, can be found here: https://sundnes.github.io/solving_odes_in_python/

calling the function inside `forward_euler`. To avoid this issue, we simply write `def f(t,u):` even if `t` is not used inside the function.[3]

For being only 15 lines of code, the capabilities of the `forward_euler` function are quite remarkable. Using this function, we can solve any kind of linear or nonlinear ODE, most of which would be impossible to solve using analytical techniques. To use this function, follow these general steps:

1. Identify $f(t, u)$ in your ODE
2. Make sure you have an initial condition u_0
3. Implement the $f(t, u)$ formula in a Python function `f(t, u)`
4. Choose the number of time steps `N`
5. Call `t, u = forward_euler(f, u0, T, N)`
6. Plot the solution

It is important to note that the FE method is the simplest of all ODE solvers, and many will argue that it is not very good. This is partly true, there exist other methods that offer greater accuracy and stability when applied to challenging ODEs. As we will see later, numerical solutions obtained using the FE method may not only be inaccurate, as depicted in Figure 1.1, but can also diverge or exhibit instability. In Chapter 3, we will explore solvers that address such issues by providing improved stability. However, the FE method is quite suitable for solving a wide range of interesting ODEs. If we are not happy with the accuracy we can simply reduce the time step, and in most cases this will give the accuracy we need with a negligible increase in computational time.

1.2 The ODE Solver Implemented as a Class

We can increase the flexibility of the `forward_euler` solver function by implementing it as a class. There are many ways to implement such a class, but one possible usage can be as follows:

```
method = ForwardEuler_v0(f)
method.set_initial_condition(u0)
t, u = method.solve(t_span=(0, 10), N=100)
plot(t, u)
```

The benefits of using a class instead of a function may not be obvious at this point, but it will become clear when we introduce different ODE solvers later. For now, let us just look at how such a solver class can be implemented to support the specified use case:

---

[3]This way of defining the right-hand side is a standard used by most available ODE solver libraries, both in Python and other languages. The right-hand side function always takes two arguments `t` and `u`, but, annoyingly, the order of the two arguments varies between different solver libraries. Some expect the `t` argument first, while others expect `u` first.

- The class should have a constructor (`__init__`) that accepts a single argument, the right-hand side function `f`, and stores it as an attribute.
- A method called `set_initial_condition` is required, which takes the initial condition as argument and stores it.
- The class should have a `solve`-method that takes the time interval `t_span` and number of time steps `N` as arguments. This method implements the for-loop for solving the ODE and returns the solution, similar to the `forward_euler` function we presented earlier.
- The time step Δt and the sequences t_n, u_n must be initialized in one of the methods, and it may also be convenient to store these as attributes. Since the time interval and the number of steps are arguments to the `solve` method, it is natural to perform these operations there.

In addition to the mentioned methods, it can be convenient to implement a separate method, for instance called `advance`, for advancing the solution one time step. This approach simplifies the implementation of new numerical methods, as we often only need to modify the `advance` method. A first version of the solver class can be implemented as follows:

```python
import numpy as np
class ForwardEuler_v0:
    def __init__(self, f):
        self.f = f

    def set_initial_condition(self, u0):
        self.u0 = u0

    def solve(self, t_span, N):
        """Compute solution for t_span[0] <= t <= t_span[1],
        using N steps."""
        t0, T = t_span
        self.dt = T / N
        self.t = np.zeros(N + 1)    # N steps ~ N+1 time points
        self.u = np.zeros(N + 1)

        msg = "Please set initial condition before calling solve"
        assert hasattr(self, "u0"), msg

        self.t[0] = t0
        self.u[0] = self.u0

        for n in range(N):
            self.n = n
            self.t[n + 1] = self.t[n] + self.dt
            self.u[n + 1] = self.advance()
        return self.t, self.u

    def advance(self):
        """Advance the solution one time step."""
        # Create local variables to get rid of "self." in
        # the numerical formula
```

```
        u, dt, f, n, t = self.u, self.dt, self.f, self.n, self.t
        return u[n] + dt * f(t[n], u[n])
```

This class performs the same tasks as the `forward_euler` function mentioned earlier, with the main advantage of the class implementation being the enhanced flexibility provided by the `advance` method. As we shall see later, implementing a different numerical method typically only requires implementing a new version of this method, leaving the rest of the code unchanged. An additional improvement in the class implementation is the inclusion of an `assert` statement within the `solve` method. This statement verifies that the user has called `set_initial_condition` before calling `solve`. Forgetting to do so is a common mistake, and the `assert` statement ensures that a useful error message is raised rather than a less informative `AttributeError`.

We can also use a class to represent the right-hand side function $f(t, u)$, which is particularly convenient for functions with parameters. Consider, for instance, the model for logistic growth:

$$u'(t) = \alpha u(t) \left(1 - \frac{u(t)}{R}\right), \quad u(0) = u_0, \quad t \in [0, 40],$$

which is typically used to model self-limiting growth of a biological population, i.e., growth that is constrained by limited resources. Initially, the growth follows an approximately exponential pattern with growth rate α. As the population size approaches the *carrying capacity* R, the population curve flattens out, see Figure 1.2 for an example solution. The right-hand side function includes two parameters α and R, but if we want to solve this model using the FE function or class, the function must be implemented as a function of t and u only. There are several ways to achieve this in Python, but a convenient approach is to implement the function as a class with a call method.[4] We can then define the parameters as attributes in the constructor and use them within the `__call__` method:

```
class Logistic:
    def __init__(self, alpha, R):
        self.alpha, self.R = alpha, float(R)

    def __call__(self, t, u):
        return self.alpha * u * (1 - u / self.R)
```

The main program for solving the logistic growth problem may now look like:

```
problem = Logistic(alpha=0.2, R=1.0)
solver = ForwardEuler_v0(problem)
u0 = 0.1
solver.set_initial_condition(u0)
t, u = solver.solve(t_span=(0, 40), N=400)
```

---

[4]Recall that if we equip a class with a special method named `__call__`, instances of the class will be callable and will behave like regular Python functions. See, for instance, Chapter 8 of [16] for a brief introduction to `__call__` and other special methods.

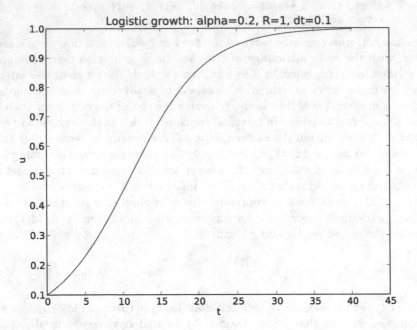

Fig. 1.2 Solution of the logistic growth model.

1.3 Systems of ODEs

Up until now, our focus has been on solving ODEs with a single solution component, commonly know as scalar ODEs. However, many interesting processes can be described by systems of ODEs, which consist of multiple ODEs where the right-hand side of one equation depends on the solution of the others. Such equation systems are also referred to as vector ODEs. One simple example is

$$u' = v, \qquad u(0) = 1$$
$$v' = -u, \quad v(0) = 0.$$

The solution of this particular system is $u = \cos t, v = \sin t$, which can be easily verified by inserting the solution into the equations and the initial conditions. For more general cases, it is usually even more difficult to find analytical solutions of ODE systems than of scalar ODEs, and numerical methods are usually necessary. In this section we will extend the solvers introduced in sections 1.1-1.2 to handle systems of ODEs. We shall see that such an extension requires relatively small modifications of the code.

Our goal is to develop general software capable of solving any vector ODE or scalar ODE. To achieve this, it is helpful to introduce some general mathematical notation. We have m unknowns

$$u^{(0)}(t), u^{(1)}(t), \ldots, u^{(m-1)}(t)$$

in a system of m ODEs:

$$\frac{d}{dt}u^{(0)} = f^{(0)}(t, u^{(0)}, u^{(1)}, \ldots, u^{(m-1)}),$$

$$\frac{d}{dt}u^{(1)} = f^{(1)}(t, u^{(0)}, u^{(1)}, \ldots, u^{(m-1)}),$$

$$\vdots = \vdots$$

$$\frac{d}{dt}u^{(m-1)} = f^{(m-1)}(t, u^{(0)}, u^{(1)}, \ldots, u^{(m-1)}).$$

To simplify the notation (and later the implementation), we can collect both the solutions $u^{(i)}(t)$ and right-hand side functions $f^{(i)}$ into vectors;

$$u = (u^{(0)}, u^{(1)}, \ldots, u^{(m-1)}),$$

and

$$f = (f^{(0)}, f^{(1)}, \ldots, f^{(m-1)}).$$

Note that f is now a vector-valued function. It takes $m+1$ input arguments (t and the m components of u) and returns a vector of m values. Using this notation, the ODE system can be written

$$u' = f(t, u), \quad u(t_0) = u_0,$$

where u and f are now vectors and u_0 is a vector of initial conditions. We observe that the notation used for scalar ODEs remains the same, and whether we are solving a scalar or system of ODEs is determined by how we define f and the initial condition u_0. This general notation is commonly employed in ODE textbooks, and we can easily make the Python implementation just as general. The use of NumPy arrays and vectorized computations greatly simplifies the generalization process and enhances the efficiency of our ODE solvers.

1.4 A `ForwardEuler` Class for Systems of ODEs

The `ForwardEuler_v0` class above was written for scalar ODEs, and we now want to modify it to handle a system of equations: $u' = f$, $u(0) = u_0$, where u, f and u_0 are vectors (arrays). To identify how the code needs to be changed, let us first revisit the underlying numerical method. Using the general notation introduced earlier, when we apply the FE method to a system of ODEs, the update formula looks exactly the same as in the scalar case, but with all the terms being vectors:

$$\underbrace{u_{k+1}}_{\text{vector}} = \underbrace{u_k}_{\text{vector}} + \Delta t \underbrace{f(u_k, t_k)}_{\text{vector}}.$$

We could also write this formula in terms of the individual components, as in

$$u_{k+1}^{(i)} = u_k^{(i)} + \Delta t f^{(i)}(t_k, u_k), \text{ for } i = 0, \ldots, m-1,$$

but the compact vector notation is more readable. Fortunately, the way we write the vector version of the formula is also how NumPy arrays are used in calculations. The Python code for the formula above may therefore look identical to the version for scalar ODEs:

```
u[k + 1] = u[k] + dt * f(t[k], u[k])
```

with the crucial difference that both `u[k]`, `u[k+1]`, and `f(t[k], u[k])` are now arrays.[5] Since these are arrays, the solution u must be a two-dimensional array, and `u[k]`,`u[k+1]`, etc. are the rows of this array. The function f expects an array as its second argument, and must return a one-dimensional array containing all the right-hand sides $f^{(0)}, \ldots, f^{(n-1)}$. To gain a better feel for how these arrays look and how they are used, let us compare the array holding the solution of a scalar ODE with that of a system of two ODEs. For the scalar equation, both t and u are one-dimensional NumPy arrays, and indexing into u gives us numbers representing the solution at each time step. For instance, in an interactive Python session we may have arrays t and u with the following contents:

```
>>> t
array([0. ,  0.4, 0.8, 1.2, ... ])
>>> u
array([1. , 1.4, 1.96, 2.744, ... ])
```

and indexing into u then gives

```
>>> u[0]
1.0
```

---

[5]This compact notation requires that the solution vector u is represented by a NumPy array. We could, in principle, use lists to hold the solution components, but the resulting code would need to loop over the components and would be far less elegant and readable.

```
>>> u[1]
1.4
```

In the case of a system of two ODEs, the array t remains one-dimensional, but the solution array u becomes two-dimensional, with one column for each solution component. We can index it in the same way as described earlier, and the result is a one-dimensional array of length two, containing the two solution components at a specific time step:

```
>>> u
array([[1.0, 0.8],
       [1.4, 1.1],
       [1.9, 2.7],
       ... ])

>>> u[0]
array([1.0, 0.8])
>>> u[1]
array([1.4, 1.1])
```

Equivalently, we could write

```
>>> u[0,:]
array([1.0, 0.8])
>>> u[1,:]
array([1.4, 1.1])
```

to make explicit which of the two array dimensions (or *axes*) that we are indexing into.

The similarity between the generic mathematical notation for vector and scalar ODEs, as well as the convenient algebra of NumPy arrays, suggests that the implementation of the solver for scalar and system ODEs can be very similar. Indeed, this is true, and the ForwardEuler_v0 class introduced earlier can be modified with a few minor adjustments to work for ODE systems:

- Ensure that f(t,u) always returns an array.
- Inspect the initial condition u0 to determine if it is a single number (scalar) or a list/array/tuple. Based on this, create the array u as either a one-dimensional or two-dimensional array.[6]

If these two aspects are handled and initialized correctly, the remaining code from Section 1.2 will work without any modifications.

The extended class implementation may look like:

```
import numpy as np

class ForwardEuler:
    def __init__(self, f):
        self.f = lambda t, u: np.asarray(f(t, u), float)
```

---

[6]This step is not strictly needed, since we could use a two-dimensional array with shape (N + 1, 1) for scalar ODEs. However, using a one-dimensional array for scalar ODEs gives simpler and more intuitive indexing.

```
    def set_initial_condition(self, u0):
        if np.isscalar(u0):                 # scalar ODE
            self.neq = 1                    # no of equations
            u0 = float(u0)
        else:                               # system of ODEs
            self.neq = u0.size              # no of equations
            u0 = np.asarray(u0)
        self.u0 = u0

    def solve(self, t_span, N):
        """Compute solution for
        t_span[0] <= t <= t_span[1],
        using N steps."""
        t0, T = t_span
        self.dt = (T - t0) / N
        self.t = np.zeros(N + 1)
        if self.neq == 1:
            self.u = np.zeros(N + 1)
        else:
            self.u = np.zeros((N + 1, self.neq))

        msg = "Please set initial condition before calling solve"
        assert hasattr(self, "u0"), msg

        self.t[0] = t0
        self.u[0] = self.u0

        for n in range(N):
            self.n = n
            self.t[n + 1] = self.t[n] + self.dt
            self.u[n + 1] = self.advance()
        return self.t, self.u

    def advance(self):
        """Advance the solution one time step."""
        u, dt, f, n, t = self.u, self.dt, self.f, self.n, self.t
        return  u[n] + dt * f(t[n], u[n])
```

It is worth commenting on certain parts of this code. First, the constructor is almost identical to the scalar case, but we use a lambda function and the convenient `np.asarray` function to convert any `f` that returns a list or tuple into a function that returns a NumPy array. If `f` already returns an array, `np.asarray` will simply return this array with no changes. This modification is not strictly necessary, since we could just assume that the user implements `f` to return an array. However, it enhances the robustness and flexibility of the class. We have also used the function `isscalar` from NumPy in the `set_initial_condition` method, to check if `u0` is a single number or a NumPy array. This allows us to determine the number of equations `self.neq` and ensures the class can handle both scalar and system ODEs. The final modification can be observed in the `solve` method, where the `self.neq` attribute is inspected. Depending on its value, `u` is initialized as a one- or two-dimensional array with the appropriate size. The actual for-loop and the `advance` method remain unchanged from the previous version of the class.

Example: ODE Model for a Pendulum. As an example, let us consider a system of ODEs that models the motion of a simple pendulum, as illustrated in Figure 1.3. This nonlinear system is a classic physics problem, and despite its simplicity, it is not possible to find an exact analytical solution. The system is formulated in terms of two main variables; the angle θ and the angular velocity ω, see Figure 1.3. For a simple pendulum with no friction, the dynamics of these variables are governed by

$$\frac{d\theta}{dt} = \omega, \tag{1.3}$$

$$\frac{d\omega}{dt} = -\frac{g}{L}\sin(\theta), \tag{1.4}$$

where L denotes the length of the pendulum and g represents the gravitational constant. Eq. (1.3) follows directly from the definition of angular velocity, while (1.4) follows from Newton's second law, where $d\omega/dt$ is the acceleration and the right-hand side is the tangential component of the gravitational force acting on the pendulum, divided by its mass. To solve the system we need to define initial conditions for θ and ω, i.e., we need to know the initial position and velocity of the pendulum.

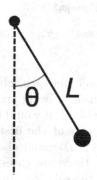

Fig. 1.3 Illustration of the pendulum problem. The main variables of interest are the angle θ and its derivative ω (the angular velocity).

Since the right-hand side defined by (1.3)-(1.4) includes the parameters L and g, it is convenient to implement it as a class, similar to the logistic growth model discussed earlier. A possible implementation may look like this:

```
from math import sin

class Pendulum:
    def __init__(self, L, g=9.81):
        self.L = L
```

```
        self.g = g

    def __call__(self, t, u):
        theta, omega = u
        dtheta = omega
        domega = -self.g / self.L * sin(theta)
        return [dtheta, domega]
```

We observe that the function returns a list. However, this list will be automatically wrapped into a function returning an array by the constructor of the solver class, as mentioned above. The main program remains quite similar to the examples presented earlier, with the exception that we now need to define an initial condition with two components. Assuming that this class definition as well as the `ForwardEuler` exist in the same file, the code to solve the pendulum problem can look like this:

```
import matplotlib.pyplot as plt

problem = Pendulum(L=1)
solver = ForwardEuler(problem)
solver.set_initial_condition([np.pi / 4, 0])
T = 10
N = 1000
t, u = solver.solve(t_span=(0, T), N=N)

plt.plot(t, u[:, 0], label=r'$\theta$')
plt.plot(t, u[:, 1], label=r'$\omega$')
plt.xlabel('t')
plt.ylabel(r'Angle ($\theta$) and angular velocity ($\omega$)')
plt.legend()
plt.show()
```

Notice that in order to extract and plot each solution component, we need to index into the second dimension of u, using array slicing. If we were to use the first index, such as u[0] or u[0, :], it would return an array of length two containing the solution components at the first time point. In this specific example, a call like plt.plot(t, u) would also work and would plot both solution components. However, there are cases where we are interested in plotting specific components of the solution, and in such cases, array slicing becomes necessary. The resulting plot is shown in Figure 1.4. Additionally, it is worth mentioning the use of Python's raw string format for the labels, indicated by the r in front of the string. Raw strings treat the backslash (\) as a regular character and are often needed when using LaTeX encoding for mathematical symbols. Furthermore, an observant reader may notice that the amplitude of the pendulum motion appears to increase over time, which is clearly not physically accurate. In reality, for an undamped pendulum problem defined by equations (1.3)-(1.4), the energy is conserved, and the amplitude should remain constant. The increasing amplitude is a numerical artifact introduced by the FE method, and the solution may be improved by reducing the time step or using a different numerical method.

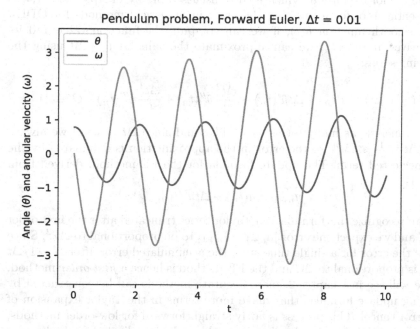

Fig. 1.4 Solution of the simple pendulum problem, computed with the forward Euler method.

1.5 Checking the Error in the Numerical Solution

Recall from Section 1.1 that we derived the FE method by approximating the derivative using a finite difference formula:

$$u'(t_n) \approx \frac{u(t_{n+1}) - u(t_n)}{\Delta t}. \tag{1.5}$$

This approximation obviously introduces an error, and since we approach the true derivative as $\Delta t \to 0$, it is intuitive that the error depends on the size of Δt. We visually demonstrated this relationship in Figure 1.1, but it would be valuable to have a way of more precisely quantifying how the error depends on the time step. Analyzing the error in numerical methods is a broad field within applied mathematics, which we will not cover in detail here, and the interested reader is referred to, for instance, [8]. However, when implementing a numerical method it is very useful to know its theoretical accuracy, and in particular to be able to compute the error and verify that the method performs as expected.

The Taylor expansion, which is also discussed briefly in Appendix A.4, is an essential tool for estimating the error in numerical methods for ODEs. For a smooth function $\hat{u}(t)$, if we can compute the function value and its derivatives at time t_n, we can approximate the value at $t_n + \Delta t$ using the following series:

$$\hat{u}(t_n + \Delta t) = \hat{u}(t_n) + \Delta t \hat{u}'(t_n) + \frac{\Delta t^2}{2}\hat{u}''(t_n) + \frac{\Delta t^3}{6}\hat{u}'''(t_n) + O(\Delta t^4).$$

We can include as many terms as we like, and since Δt is small we always have $\Delta t^{(n+1)} \ll \Delta t^n$, so the error in the approximation is dominated by the first neglected term. The update formula for the FE method, derived from (1.5), is

$$u_{n+1} = u(t_n) + \Delta t u'(t_n),$$

We can recognize this formula as a Taylor series truncated after the first order term, and we expect the error $|u_{n+1} - \hat{u}_{n+1}|$ to be proportional to Δx^2. Since this is the error for a single time step, the accumulated error after $N \sim 1/\Delta t$ steps is proportional to Δt, and the FE method is hence a *first order* method. As we will see in Chapter 2, more accurate methods can be constructed by deriving update formulas that make more terms in the Taylor expansion of the error cancel. This process is fairly straightforward for low-order methods, e.g., of second or third order, but it quickly gets complicated for high order solvers, see, for instance [8] for details.

Knowing the theoretical accuracy of an ODE solver is important for a number of reasons, including the verification of the solver implementation. If we can solve a given problem and demonstrate that the error behaves as predicted by the theory, we gain confidence in the correctness of our solver. To illustrate this procedure, let us consider the simple initial value problem introduced earlier:

$$u' = u, \quad u(0) = 1.$$

As stated above, the analytical solution to this problem is $u = e^t$, and we can use this to compute the error in our numerical solution. But how should the error be defined? There is no unique answer to this question. For practical applications, common error measures include the root-mean-square (RMS) or relative-root-mean-square (RRMS), which are defined by

$$RMSE = \sqrt{\frac{1}{N}\sum_{n=0}^{N}(u_n - \hat{u}(t_n))^2},$$

$$RRMSE = \sqrt{\frac{1}{N}\sum_{n=0}^{N}\frac{(u_n - \hat{u}(t_n))^2}{\hat{u}(t_n)^2}},$$

respectively. Here, u_n is the numerical solution at time step n and $\hat{u}(t_n)$ the corresponding exact solution. In more mathematically oriented texts, the errors are usually defined in terms of *norms*, for instance the discrete l_1, l_2, and l_∞ norms:

$$e_{l_1} = \sum_{i=0}^{N}(|u_i - \hat{u}(t_i)|),$$

$$e_{l_2} = \sum_{i=0}^{N}(u_i - \hat{u}(t_i)^2),$$

$$e_{l_\infty} = \max_{i=0}^{N}(\hat{u}_i - u(t_i)).$$

While the choice of error norm may be important for certain cases, it is usually not crucial for practical applications, and all the different error measures can generally be expected to behave as predicted by the theory. For simplicity, we will use an even simpler error measure in our example, where we compute the error at the final time T, given by $e = |u_N - \hat{u}(t_N)|$. Using the `ForwardEuler` class introduced above, the complete code for checking the convergence can be written as follows:

```
from forward_euler_class_v1 import ForwardEuler
import numpy as np

def rhs(t, u):
    return u

def exact(t):
    return np.exp(t)

solver = ForwardEuler(rhs)
solver.set_initial_condition(1.0)

T = 3.0
t_span = (0,T)
N = 30

print('Time step (dt)   Error (e)      e/dt')
for _ in range(10):
    t, u = solver.solve(t_span, N)
    dt = T / N
    e = abs(u[-1] - exact(T))
    print(f'{dt:<14.7f}   {e:<12.7f}   {e/dt:5.4f}')
    N = N * 2
```

Most of the lines in the code are identical to the previous programs. However, we have enclosed the call to the `solve` method within a for loop, and the last line ensures that the number of time steps N is doubled for each iteration of the loop. Also, note the f-string format specifiers used, such as `{dt:<14.7f}`, which specifies that the output should be a left-aligned decimal number with seven decimals, occupying a total of 14 characters. These format specifiers ensure that the numbers are displayed as vertically aligned columns, improving readability, which may be important for visually inspecting the convergence. See, for instance, [16] for a brief introduction to f-strings and format specifiers. The program will produce the following output:

```
Time step (dt)    Error (e)        e/dt
0.1000000         2.6361347        26.3613
0.0500000         1.4063510        28.1270
0.0250000         0.7273871        29.0955
0.0125000         0.3700434        29.6035
0.0062500         0.1866483        29.8637
0.0031250         0.0937359        29.9955
0.0015625         0.0469715        30.0618
0.0007813         0.0235117        30.0950
0.0003906         0.0117624        30.1116
0.0001953         0.0058828        30.1200
```

In the rightmost column we observe that the ratio of the error to the time step remains approximately constant. This observation supports the theoretical result that the error is proportional to Δt. In the upcoming chapters, we will perform similar calculations for higher order methods to verify that the error is proportional to Δt^r, where r is the theoretical order of convergence for the method.

To compute the error in our numerical solution, we need to determine the true solution to our initial value problem. This task was straightforward for the simple example above because we knew the analytical solution to the equation. However, for more complex ODE problems, estimating the error and the order of convergence requires a different approach. Several alternatives exist, including methods like the *method of manufactured solutions*, where we choose a solution function $u(t)$ and compute its derivative analytically to determine the right-hand side of the ODE. An even simpler approach, which usually yields good results, involves computing a highly accurate numerical solution using a high-order solver and small time steps. This solution then serves as the reference for computing the error. To obtain accurate error estimates it is essential that the reference solution is significantly more accurate than the numerical solution we want to evaluate. Generating the reference solution typically requires very small time steps and can take some time to compute, but in most cases the computation time for the reference solution is not a significant issue.

1.6 Using ODE Solvers from SciPy

As mentioned in the book's preface, there exist many ODE solvers available for direct use, and it can be argued that there is no need to implement our own solvers. While there is some truth to this, as we have emphasized, it can be beneficial to understand the inner workings of these solvers, in order to apply and use them correctly, and the best way to obtain this knowledge is to implement the solvers ourselves. However, when we have a specific ODE model that we need to solve as efficiently as possible, there are several existing solvers to choose from. For Python programmers, a natural choice may be the solvers provided by *SciPy*, which have evolved into a robust and fairly efficient suite of ODE solvers. SciPy is a comprehensive scientific software package for Python that includes various tools for tasks such as linear algebra, optimization, integration, and more.[7] To solve initial value problems, the recommended tool is the `solve_ivp` function from the `integrate` module in SciPy. The following code demonstrates the use of `solve_ivp` with the `Pendulum` class presented above to solve the simple pendulum problem defined by (1.3)-(1.4). We here assume that the `Pendulum` class is stored in a separate file named `pendulum.py`.

```
from scipy.integrate import solve_ivp
import numpy as np
import matplotlib.pyplot as plt
from pendulum import Pendulum

problem = Pendulum(L = 1)
t_span = (0, 10.0)
u0 = (np.pi/4, 0)

solution = solve_ivp(problem, t_span, u0)

plt.plot(solution.t, solution.y[0,:])
plt.plot(solution.t, solution.y[1,:])
plt.legend([r'$\theta$',r'$\omega$'])
plt.show()
```

Running this code will generate a plot similar to Figure 1.5, and we observe that the solution does not appear as smooth as the one obtained from the `ForwardEuler` solver introduced earlier. This seeming discrepency is due to the nature of the `solve_ivp` solver, which is an *adaptive* solver that automatically selects the time step to meet a specified error tolerance. The default value of this tolerance is relatively large, leading to the solver using very few time steps and resulting in jagged-looking solution plots. Comparing the plot with a highly accurate numerical solution, represented by the two dotted curves in Figure 1.5, we notice that the solution at the specified time points t_n is fairly accurate. However, the visual appearance is compromised by the linear interpolation between these time points. To obtain a more visually appealing solution, there are several approaches we can take. We may, for instance, pass the function an additional argument `t_eval`, which is a NumPy array containing the desired time points for evaluating the solution:

```
t_eval = np.linspace(0, 10.0, 1001)
```

---

[7] See https://scipy.org/

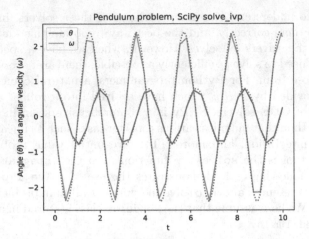

Fig. 1.5 Solution of the simple pendulum problem, computed with the SciPy `solve_ivp` function and the default tolerance.

```
solution = solve_ivp(problem, t_span, u0, t_eval=t_eval)
```

Alternatively, we can reduce the error tolerance of the solver, for instance, by setting

```
rtol = 1e-6
solution = solve_ivp(problem, t_span, u0, rtol=rtol)
```

This latter call will reduce the relative tolerance `rtol` from its default value of `1e-3` (0.001). We could also adjust the absolute tolerance using the parameter `atol`. We will not cover all the possible arguments and options to the `solve_ivp` function here, but it is worth mentioning that we can also change the numerical method used by the function, by passing in a parameter named `method`. For instance, a call like

```
rtol = 1e-6
solution = solve_ivp(problem, t_span, u0, method='Radau')
```

will replace the default solver (called `rk45`) with an implicit Radau ODE solver, which we will cover in Chapter 3. For a complete description of parameters accepted by the `solve_ivp` function we recommend referring to the online documentation available on the SciPy website.

Chapter 2
Improving the Accuracy

As mentioned previously, the FE method derived in Chapter 1 is not the most sophisticated ODE solver. Although it provides sufficient accuracy for most of the applications covered in this book, there are alternative methods available that offer improved accuracy and stability, making them better suited for solving challenging ODE systems. In this chapter, we will focus on enhancing accuracy, and thus we will primarily explore explicit methods. Implicit methods, which exhibit superior stability properties and are better suited for solving *stiff* ODEs, will be discussed in Chapter 3.

In Chapter 1 we demonstrated that the FE method is a first-order accurate method, meaning that the error in the numerical solution is proportional to the size of the time step Δt. In this chapter, we will derive solvers of higher order, where the numerical error is proportional to a higher power of Δt. To explain the derivation of such higher order ODE solvers, we will revisit the general formulation of the ODE system:

$$u' = f(t, u), \quad u(t_0) = u_0.$$

Instead of simply replacing the derivative with a finite difference approximation, as done in the derivation of the FE method, we will adopt a slightly different approach in this chapter. Assuming that we know the solution u_n at time t_n, we can find the solution at time t_{n+1} by integrating both sides of the equation. We have

$$\int_{t_n}^{t_{n+1}} \frac{du}{dt} dt = \int_{t_n}^{t_{n+1}} f(t, u(t)) dt,$$

which gives us the exact solution at time t_{n+1} as

$$u(t_{n+1}) = u(t_n) + \int_{t_n}^{t_{n+1}} f(t, u(t)) dt. \tag{2.1}$$

© The Author(s) 2024
J. Sundnes, *Solving Ordinary Differential Equations in Python*,
Simula SpringerBriefs on Computing 15,
https://doi.org/10.1007/978-3-031-46768-4_2

In general, it is not feasible to compute the integral on the right-hand side of the equation analytically, as the function f is often nonlinear and the function $u(t)$ is unknown. However, we can approximate the integral using various numerical integration techniques. The simplest approximation is to set

$$f(t, u(t)) \approx f(t_n, u_n), \text{ for } t_n < t < t_{n+1},$$

meaning that we approximate the integrand as a constant over the interval from t_n to $t_n + \Delta t$. Substituting this choice into (2.1) yields the update formula:

$$u_{n+1} = u_n + \Delta t \, f(t_n, u_n),$$

which is recognized as the FE method introduced in Chapter 1. Approximating the function $f(t_n, u_n)$ as constant over the interval $t_n < t < t_{n+1}$ is obviously not the most accurate choice, and we will explore more accurate approximations that lead to higher-order ODE solvers.

The classical approach for approximating the integral of a general nonlinear function is to approximate the function with a polynomial and then integrate this polynomial analytically. This technique forms the basis for standard quadrature rules in numerical integration, and has also been used to derive accurate ODE solvers. Two main ideas have been explored for constructing the polynomial approximation of $f(t, u)$, resulting in two important classes of ODE solvers. The first approach is to approximate $f(t, u)$ with a polynomial that interpolates f at previous time points, such as $f(t_{n-1}, u_{n-1}), f(t_{n-2}, u_{n-2}), \ldots$ This method gives rise to *multistep* methods, which are widely used for solving ODEs. We will not cover multistep methods in this book, but interested readers can refer to references [1, 8] for further details. The second approach entails computing a number of intermediate approximations of $f(t, u)$ on the interval $t_n < t < t_{n+1}$, and using these values to define the polynomial approximation of f. This approach is analogous to the derivation of classical quadrature rules in numerical integration, and leads to a class of ODE solvers known as Runge-Kutta methods. These methods come in many forms, exhibiting different accuracy and stability properties, and will be the main focus of Chapters 2-4.

2.1 Explicit Runge-Kutta Methods

As mentioned earlier, an intuitive way to improve the accuracy of the approximate integral in (2.1) is to calculate several intermediate approximations of $f(t_*, u_*)$ for $t_n \leq t_* \leq t_{n+1}$, and calculate the integral as a weighted sum of these values. This approach builds upon standard numerical integration techniques and gives rise to a widely used class of ODE solvers called *Runge-Kutta (RK) methods*. The simplest example of an RK method is in fact the FE method discussed earlier, which is a one-stage, first-order, explicit RK method. An alternative formulation of the FE method is

$$k_1 = f(t_n, u_n),$$
$$u_{n+1} = u_n + \Delta t k_1.$$

It can be observed that this is the same formula as introduced earlier, and there is no real advantage in writing the formula in two lines instead of one. However, this alternative formulation aligns with the typical representation of RK methods and facilitates understanding the relationship between the FE method and more advanced solvers. The intermediate value k_1 is often referred to as a *stage derivative* in the ODE literature.

To enhance the accuracy of the FE method to second order, i.e., with error proportional to Δt^2, we can employ more accurate approximations of the integral in (2.1). One option is to maintain the assumption that $f(t, u(t))$ is constant over $t_n \leq t_* \leq t_{n+1}$, but to approximate it at the midpoint of the interval instead of the left end. This approach requires one additional stage:

$$k_1 = f(t_n, u_n), \tag{2.2}$$
$$k_2 = f(t_n + \frac{\Delta t}{2}, u_n + \frac{\Delta t}{2} k_1), \tag{2.3}$$
$$u_{n+1} = u_n + \Delta t k_2. \tag{2.4}$$

This method is known as the explicit midpoint method or the modified Euler method. The first step is identical to that of the FE method, but instead of using the stage derivative k_1 to advance the solution to the next step, we use it to calculate an intermediate midpoint solution

$$u_{n+1/2} = u_n + \frac{\Delta t}{2} k_1.$$

This solution is then used to compute the corresponding stage derivative k_2, which serves as an approximation to the derivative of u at time $t_n + \Delta t/2$. Finally, we use this midpoint derivative to advance the solution to t_{n+1}.

Another second-order method is Heun's method, also known as the explicit trapezoidal method, which can be derived by approximating the integral in equation (2.1) using the trapezoidal rule:

$$k_1 = f(t_n, u_n), \tag{2.5}$$
$$k_2 = f(t_n + \Delta t, u_n + \Delta t k_1), \tag{2.6}$$
$$u_{n+1} = u_n + \frac{\Delta t}{2}(k_1 + k_2). \tag{2.7}$$

This method also computes two stage derivatives k_1 and k_2. However, note that the formula for k_2 approximates the derivative at t_{n+1} rather than at the midpoint $t_n + \Delta t/2$. The solution is then advanced from t_n to t_{n+1} using the mean value of k_1 and k_2.

All RK methods follow the same recipe as the two second-order methods considered above; we calculate one or more intermediate values (i.e., stage derivatives) and then advance the solution using a combination of these stage derivatives. The method's accuracy can be improved by adding more stages. A general RK method with s stages can be written as

$$k_i = f(t_n + c_i \Delta t, u_n + \Delta t \sum_{j=1}^{s} a_{ij} k_j), \text{ for } i = 1, \ldots, s \qquad (2.8)$$

$$u_{n+1} = u_n + \Delta t \sum_{i=1}^{s} b_i k_i. \qquad (2.9)$$

Here c_i, b_i, a_{ij}, for $i, j, = 1, \ldots, s$ are coefficients specific to the method. Every RK method can be written in this form, and a method is uniquely determined by the number of stages, s, and the values of the coefficients.

As mentioned earlier, there exists a wide variety of RK methods, where the coefficients are typically chosen to optimize the accuracy for a given number of stages. While we will not delve into the details of how the methods are constructed here, it can be useful to mention some of the underlying principles. For instance, it can be shown that the b_i coefficients must satisfy $\sum_{i=1}^{s} b_i = 1$ in order to ensure convergence. This condition naturally arises from the motivation for RK methods as numerical integrators applied to equation (2.1). When approximating the integral as a weighted sum, the sum of the weights must be equal to one. Another common constraint on the coefficients is to set $c_i = \sum_{j=1}^{s} a_{ij}$. While not strictly necessary, this constraint can simplify the derivation of the methods and aligns with our interpretation of the stage derivative k_i as approximations of the right-hand side $f(t, u)$ at time $t_n + c_i \Delta t$. When implementing a new solver it is easy to introduce errors in the coefficient values, and it may be useful to include tests to verify that the most fundamental conditions on the coefficients are satisfied. It is possible to derive general *order conditions* that the coefficients must satisfy for a method to achieve a given order, see, for instance, [1,8] for details. In this chapter we exclusively consider explicit Runge-Kutta (ERK) methods, which means that $a_{ij} = 0$ for $j \geq i$. It can be shown that the order p of an ERK method with s stages satisfies $p \leq s$, and for $p \geq 5$, the bound is $p \leq s - 1$. However, it remains unknown whether this latter bound is sharp, and it may be even stricter for methods of very high order. For instance, all known methods with $p = 8$ have at least eleven stages, and it is not known whether eight-order methods with nine or ten stages exist.

In the ODE literature, method coefficients are often specified in the form of a *Butcher tableau*, which provides a concise representation of any RK method. The Butcher tableau is simply a specification of all the method coefficients, and for a general RK method it is written as

$$\begin{array}{c|ccc} c_i & a_{11} & \cdots & a_{1s} \\ \vdots & \vdots & \vdots & \vdots \\ c_s & a_{s1} & \cdots & a_{ss} \\ \hline & b_1 & \cdots & b_s \end{array}$$

The Butcher tableaus of the three methods discussed above: FE, explicit midpoint, and Heun's method, are

$$\begin{array}{c|c} 0 & 0 \\ \hline & 1 \end{array}, \qquad \begin{array}{c|cc} 0 & 0 & 0 \\ 1/2 & 1/2 & 0 \\ \hline & 0 & 1 \end{array}, \qquad \begin{array}{c|cc} 0 & 0 & 0 \\ 1 & 1 & 0 \\ \hline & 1/2 & 1/2 \end{array},$$

respectively. To grasp the concept of Butcher tableaus, it is beneficial to practice inserting the coefficients from these three tableaus into equations (2.8)-(2.9) and verifying that the correct formulas for the respective methods are obtained. As an example of a higher order method, we may consider the "original" Runge-Kutta method, which is a fourth-order, four-stage method defined by

$$\begin{array}{c|cccc} 0 & 0 & 0 & 0 & 0 \\ 1/2 & 1/2 & 0 & 0 & 0 \\ 1/2 & 0 & 1/2 & 0 & 0 \\ 1 & 0 & 0 & 1 & 0 \\ \hline & 1/6 & 1/3 & 1/3 & 1/6 \end{array},$$

which gives the formulas

$$k_1 = f(t_n, u_n), \tag{2.10}$$

$$k_2 = f(t_n + \frac{\Delta t}{2}, u_n + \frac{\Delta t}{2} k_1), \tag{2.11}$$

$$k_3 = f(t_n + \frac{\Delta t}{2}, u_n + \frac{\Delta t}{2} k_2), \tag{2.12}$$

$$k_4 = f(t_n + \Delta t, u_n + \Delta t k_3), \tag{2.13}$$

$$u_{n+1} = u_n + \frac{\Delta t}{6} (k_1 + 2k_2 + 2k_3 + k_4). \tag{2.14}$$

As mentioned earlier, all the methods discussed in this chapter are explicit methods, meaning that $a_{ij} = 0$ for $j \geq i$. Examining equations (2.10)-(2.14) or the general formula (2.8) more closely, we observe that this conditions implies that each stage derivative k_i only depends on previously computed stage derivatives. Consequently, all k_i can be computed sequentially using explicit formulas. In contrast, for implicit RK methods, $a_{ij} \neq 0$ for some $j \geq i$. As seen in equation (2.8), the formula for computing k_i will then include k_i on the right-hand side, as part of the argument to the function f. Therefore, equations need to be solved to compute the stage derivatives, and since f is typically nonlinear, we need to solve these equations with an iterative solver such as Newton's method. These steps make implicit RK methods more complex to implement and more computationally expensive per time step, but they are also more stable than explicit methods and perform much better for certain classes of ODEs. We will cover implicit RK methods in Chapter 3.

2.2 A Class Hierarchy of Runge-Kutta Methods

We now want to implement RK methods as classes, similar to the FE classes introduced above. Upon examining the `ForwardEuler` class, we may notice that most of the code is common to all ODE solvers and is not specific to the FE method. For instance, we always need to create an array to store the solution, and the general solution method using a for-loop remains the same for all methods. The only difference among the methods lies in how the solution is advanced from one time step to the next. Recalling the ideas of Object-Oriented Programming, it becomes apparent that a class hierarchy is a suitable structure for implementing such a collection of ODE solvers. This approach allows us to consolidate common code in a superclass (base class), and use inheritance to avoid code duplication. The superclass can handle most of the administrative steps of the ODE solver, such as

- Storing the solution u_n and the time points t_n, $k = 0, 1, 2, \ldots, n$
- Storing the right-hand side function $f(t, u)$
- Storing and applying the initial condition
- Running the loop over all time steps

To address these aspects, we can introduce a superclass called `ODESolver` and implement the method-specific details in subclasses. This is precisely why we isolated the code to perform a single step in the `advance` method, as it becomes the only method that needs to be implemented in the subclasses. The implementation of the superclass can closely resemble the `ForwardEuler` class introduced earlier:

```python
import numpy as np

class ODESolver:
    def __init__(self, f):
        # Wrap user's f in a new function that always
        # converts list/tuple to array (or let array be array)
        self.model = f
        self.f = lambda t, u: np.asarray(f(t, u), float)

    def set_initial_condition(self, u0):
        if np.isscalar(u0):            # scalar ODE
            self.neq = 1               # no of equations
            u0 = float(u0)
        else:                          # system of ODEs
            u0 = np.asarray(u0)
```

```
            self.neq = u0.size          # no of equations
        self.u0 = u0

    def solve(self, t_span, N):
        """Compute solution for t_span[0] <= t <= t_span[1],
        using N steps."""
        t0, T = t_span
        self.dt = (T - t0) / N
        self.t = np.zeros(N + 1)   # N steps ~ N+1 time points
        if self.neq == 1:
            self.u = np.zeros(N + 1)
        else:
            self.u = np.zeros((N + 1, self.neq))

        msg = "Please set initial condition before calling solve"
        assert hasattr(self, "u0"), msg

        self.t[0] = t0
        self.u[0] = self.u0

        for n in range(N):
            self.n = n
            self.t[n + 1] = self.t[n] + self.dt
            self.u[n + 1] = self.advance()
        return self.t, self.u

    def advance(self):
        raise NotImplementedError(
            "Advance method is not implemented in the base class")
```

It is important to note that the ODESolver is designed to be a pure superclass, and the implementation of the advance method is left for subclasses. To clearly convey this abstract nature of the class, we have included an advance method that raises a NotImplementedError when it is called. If an attempt is made to create an instance of ODESolver and use it as a standalone solver, an error will occur in the line self.u[n + 1] = self.advance(). Omitting the definition of advance entirely would lead to an error in the same line, but in this case it would be a less informative AttributeError. By raising the NotImplementedError, it explicitly indicates to anyone reading or using the code that this behavior is intentional and that the specific functionality is meant to be implemented in subclasses. It is worth mentioning that there are alternative approaches in Python to make explicit the abstract nature of the ODESolver class, for instance using the module abc, for "Abstract Base Class". However, while this solution may be considered more modern, we have decided to not use it here, in the interest of keeping the code simple and compact.

With the superclass at hand, the implementation of a ForwardEuler subclass becomes very simple:

```
class ForwardEuler(ODESolver):
    def advance(self):
```

```
        u, f, n, t = self.u, self.f, self.n, self.t
        dt = self.dt
        return u[n] + dt * f(t[n], u[n])
```

Similarly, the explicit midpoint method and the fourth-order RK method can be subclasses, each implementing a single method:

```
class ExplicitMidpoint(ODESolver):
    def advance(self):
        u, f, n, t = self.u, self.f, self.n, self.t
        dt = self.dt
        dt2 = dt / 2.0
        k1 = f(t[n], u[n])
        k2 = f(t[n] + dt2, u[n] + dt2 * k1)
        return u[n] + dt * k2

class RungeKutta4(ODESolver):
    def advance(self):
        u, f, n, t = self.u, self.f, self.n, self.t
        dt = self.dt
        dt2 = dt / 2.0
        k1 = f(t[n], u[n],)
        k2 = f(t[n] + dt2, u[n] + dt2 * k1, )
        k3 = f(t[n] + dt2, u[n] + dt2 * k2, )
        k4 = f(t[n] + dt, u[n] + dt * k3, )
        return u[n] + (dt / 6.0) * (k1 + 2 * k2 + 2 * k3 + k4)
```

The use of these classes is nearly identical to the FE class introduced in Section 1.3. Considering the same simple ODE used above; $u' = u$, $u(0) = 1$, $t \in [0,3]$, $\Delta t = 0.5$, the code looks like:

```
import numpy as np
import matplotlib.pyplot as plt
from ODESolver import ForwardEuler, ExplicitMidpoint, RungeKutta4

def f(t, u):
    return u

t_span = (0, 3)
N = 6

fe = ForwardEuler(f)
fe.set_initial_condition(u0=1)
t1, u1 = fe.solve(t_span, N)
plt.plot(t1, u1, label='Forward Euler')

em = ExplicitMidpoint(f)
em.set_initial_condition(u0=1)
t2, u2 = em.solve(t_span, N)
plt.plot(t2, u2, label='Explicit Midpoint')

rk4 = RungeKutta4(f)
rk4.set_initial_condition(u0=1)
t3, u3 = rk4.solve(t_span, N)
```

```
plt.plot(t3, u3, label='Runge-Kutta 4')

# plot the exact solution in the same plot
time_exact = np.linspace(0, 3, 301)
plt.plot(time_exact, np.exp(time_exact), label='Exact')
plt.title('RK solvers for exponential growth, $\\Delta t = 0.5$')
plt.xlabel('$t$')
plt.ylabel('$u(t)$')
plt.legend()
plt.show()
```

This code will solve the same simple equation using three different methods, and plot the solutions in the same window, as shown in Figure 2.1. To emphasize the disparity in accuracy between the methods, we have set N = 6, resulting in a very large time step ($\Delta t = 0.5$).

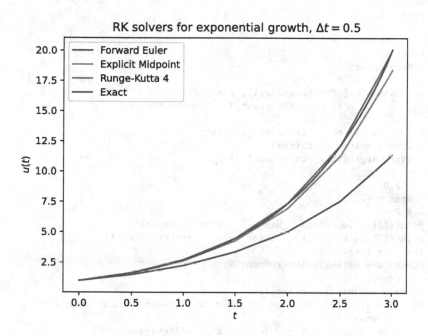

Fig. 2.1 Numerical solutions of the exponential growth problem, computed with ForwardEuler, ImplicitMidpoint and RungeKutta4. All the solvers use $\Delta t = 0.5$, to highlight the difference in accuracy.

2.3 Testing the Solvers

In Chapter 1 we demonstrated how to compute the error in the numerical solution, and in particular how we could verify that the error behaved as predicted by the theoretical convergence of the applied solvers. Such tests are extremely valuable for verifying that we have implemented the ODE solvers correctly, and can easily be extended to the higher order solvers. As an example, the following code defines a dictionary containing three different solver classes and their theoretical order, and solves the simple exponential ODE using all three solvers.

```python
from ODESolver import *
import numpy as np


def rhs(t, u):
    return u


def exact(t):
    return np.exp(t)

solver_classes = [(ForwardEuler,1), (Heun,2),
                  (ExplicitMidpoint,2), (RungeKutta4,4)]

for solver_class, order in solver_classes:
    solver = solver_class(rhs)
    solver.set_initial_condition(1.0)

    T = 3.0
    t_span = (0, T)
    N = 30
    print(f'{solver_class.__name__}, order = {order}')
    print(f'Time step (dt)    Error (e)      e/dt**{order}')
    for _ in range(10):
        t, u = solver.solve(t_span, N)
        dt = T / N
        e = abs(u[-1] - exact(T))
        if e < 1e-13:  # break if error is close to machine precision
            break
        print(f'{dt:<14.7f}  {e:<12.7f}  {e/dt**order:5.4f}')
        N = N * 2
```

The code is nearly identical to the FE convergence test in Section 1.5, with the only difference being that we loop over a list of tuples containing the four method classes and their corresponding orders. The output is also similar to the previous version, but now repeated for all four solvers. The built-in class attribute __name__ is used to extract and print the name of each solver. Three columns are displayed, representing the time step Δt, the error e at time $t = 3.0$, and finally $e/\Delta t^p$, where p is the order of the method. The output matches the expected values for the first two methods, as the numbers in the rightmost column are approximately constant, confirming that the error is in fact proportional to Δt^p. However, the last part of the output, for the forth order RK method, looks like this:

```
RungeKutta4 order = 4
Time step (dt)    Error (e)      e/dt**4
0.1000000         0.0000462      0.4620
0.0500000         0.0000030      0.4817
0.0250000         0.0000002      0.4918
0.0125000         0.0000000      0.4969
0.0062500         0.0000000      0.4995
0.0031250         0.0000000      0.5006
0.0015625         0.0000000      0.5025
0.0007813         0.0000000      0.5436
0.0003906         0.0000000      5.1880
0.0001953         0.0000000      102.5391
```

We see that the $e/\Delta t^p$ numbers remain approximately constant for a while, consistent with the convergence order of the method. However, for very small values of Δt, these values start to increase. This behavior is not uncommon in convergence tests, especially for high-order methods, and is caused by the finite accuracy of number representation on computers. As the numerical errors become smaller and approach the machine precision ($\approx 10^{-16}$), roundoff errors start to dominate the overall error, leading to a loss of convergence.

There are many alternative ways to check the implementation of ODE solvers. One approach is to consider an even simpler ODE where the right-hand side is a constant, i.e., $u'(t) = f(t, u) = C$. The solution to this simple ODE is given by $u(t) = Ct + u_0$, where u_0 is the initial condition. All the numerical methods discussed in this book will capture this solution to machine precision, and we can write a general test function that takes advantage of this:

```python
def test_exact_numerical_solution():
    solver_classes = [ForwardEuler, Heun,
                      ExplicitMidpoint, RungeKutta4]
    a = 0.2
    b = 3

    def f(t, u):
        return a

    def u_exact(t):
        """Exact u(t) corresponding to f above."""
        return a * t + b

    u0 = u_exact(0)
    T = 8
    N = 10
    tol = 1E-14
    t_span = (0, T)
    for solver_class in solver_classes:
        solver = solver_class(f)
```

```
solver.set_initial_condition(u0)
t, u = solver.solve(t_span, N)
u_e = u_exact(t)
max_error = abs((u_e - u)).max()
msg = f'{solver_class.__name__} failed, error={max_error}'
assert max_error < tol, msg
```

Similar to the convergence check illustrated below, this code will loop through all the solver classes, solve the simple ODE, and check that the resulting error falls within the specified tolerance.

Both of the methods shown here for verifying the implementation of our solvers have certain limitations. The most important one is that they both solve very simple ODEs, and it is possible to introduce errors in the code that may only manifest themselves when dealing with more complex problems. However, the methods presented here offer the advantages of simplicity and generality, and they can be applied to any newly implemented ODE solver class. Many common implementation errors, such as incorrectly specifying a single parameter in an RK method, will often become apparent even when solving these simple problems. Therefore, these methods can provide an initial indication of whether the implementation is correct, which can be followed by more extensive tests if needed.

Chapter 3
Stable Solvers for Stiff ODE Systems

In the previous chapter, we introduced explicit Runge-Kutta (ERK) methods and demonstrated how they can be implemented as a hierarchy of Python classes. For most ODE systems, replacing the simple forward Euler method with a higher-order ERK method will significantly reduce the number of time steps needed to reach a specified accuracy. Furthermore, it often leads to reduced computation time, since the additional cost per time step is outweighed by the reduced number of steps. However, there exists a class of ODEs known as *stiff* systems, where all the ERK methods require very small time steps, and any attempt to increase the time step leads to spurious oscillations and possible divergence of the solution. Stiff ODE systems pose a challenge for explicit methods, and they are better addressed by implicit solvers such as implicit Runge-Kutta (IRK) methods. IRK methods are well-suited for stiff problems and can offer substantial reductions in computation time when tackling challenging problems.

3.1 Stiff ODE Systems and Stability

One well-known example of a stiff ODE system is the Van der Pol equation, which can be written as an initial value problem on the form

$$y_1' = y_2, \qquad\qquad y_1(0) = 1, \qquad\qquad (3.1)$$
$$y_2' = \mu(1 - y_1^2)y_2 - y_1, \quad y_2(0) = 0. \qquad\qquad (3.2)$$

Here, the parameter μ represents a constant that determines the properties of the system, including its "stiffness". When $\mu = 0$ the problem is a simple oscillator with analytical solution $y_1 = \cos(t), y_2 = \sin(t)$. However, for non-zero values of μ, the solution exhibits far more complex behavior. The following code implements this system and solves it with the `ForwardEuler` subclass from the `ODESolver` class hierarchy.

© The Author(s) 2024
J. Sundnes, *Solving Ordinary Differential Equations in Python*,
Simula SpringerBriefs on Computing 15,
https://doi.org/10.1007/978-3-031-46768-4_3

```
from ODESolver import *
import numpy as np
import matplotlib.pyplot as plt

class VanderPol:
    def __init__(self, mu):
        self.mu = mu

    def __call__(self, t, u):
        du0 = u[1]
        du1 = self.mu * (1 - u[0]**2) * u[1] - u[0]
        return du0, du1

model = VanderPol(mu=1)

solver = ForwardEuler(model)
solver.set_initial_condition([1, 0])

t,u  = solver.solve(t_span=(0, 20), N=1000)

plt.plot(t, u)
plt.show()
```

Figure 3.1 shows the solutions of the Van der Pol equation for $\mu = 0, 1$ and 5. When the parameter μ is set even higher, such as $\mu = 50$, the solution diverges (becomes unstable) with the given time step ($\Delta t = 0.02$). Although using a more accurate ERK method instead of the FE method may provide some improvement, it does not resolve the issue significantly. It does help to reduce the time step considerably, but the resulting computation time may be substantial. In this problem, the time step is determined by stability requirements rather than the desired accuracy, and opting for a solver that is more stable than the previously discussed ERK methods may yield significant benefits.

Before introducing more stable solvers, it is useful to examine the observed stability problems in more detail. Why does the solution of the Van der Pol model deteriorate significantly for large values of μ? More generally, what are the properties of an ODE system that make it stiff? To address these questions, it is useful to start with a simpler problem than the Van der Pol model. Consider, for instance, a simple IVP known as the Dahlquist test equation:

$$u' = \lambda u, \quad u(0) = 1, \tag{3.3}$$

where λ can be a complex number.[1] When $\lambda = 1$, it corresponds to the simple exponential growth problem we discussed earlier. However, in this chapter

---

[1]Note that the implementation of the solvers in this book does not support solving this ODE for complex λ. However, considering complex values in the stability analysis is still important because, for systems of ODEs, the relevant values are the *eigenvalues* of the right-hand side, and these may be complex.

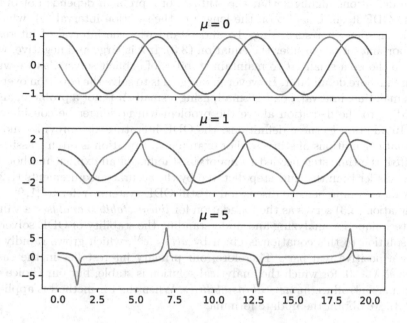

Fig. 3.1 Solutions of the Van der Pol model for different values of μ.

we primarily focus on λ values with a negative real part, i.e., either real or complex λ values that satisfy $\Re(\lambda) < 0$. In such cases, the solution of equation (3.3) decays over time and remains stable. However, we will discover that the numerical solutions may not always preserve this stability.

Following the definition in [1], we classify problem (3.3) as stiff for an interval $[0, b]$ if the real part of λ satisfies

$$b\Re(\lambda) \ll -1.$$

For more general nonlinear problems, such as the Van der Pol model in (3.1)-(3.2), the stiffness of the system is determined by the eigenvalues λ_i of the local Jacobian matrix J, which is the matrix of partial derivatives of the right-hand side function f. The Jacobian is defined by

$$J_{ij} = \frac{\partial f_i(t, y)}{\partial y_j},$$

and the problem is considered stiff for an interval $[0, b]$ if

$$b\min_i \Re(\lambda_i) \ll -1.$$

These definitions highlight that the stiffness of a problem depends not only on the ODE itself, but also on the length of the solution interval (b), which may seem somewhat surprising. To understand why the interval of interest is important, let us consider the equation (3.3). If λ is large and negative, we need to choose a small Δt to maintain stability of explicit solvers, as we will discuss in more detail later. However, if our goal is to solve the equation over a very small time interval, i.e., b is small, using a small Δt is not a problem, and according to the definition above the problem will no longer be considered stiff. In addition to these definitions, the ODE literature also provides more pragmatic definitions of stiffness. For example, an equation is often classified as stiff if the time step needed to maintain stability of an explicit method is much smaller than the time step dictated by the accuracy requirements [1,2]. For a more comprehensive discussion of stiff ODE systems, refer to [1,9].

Equation (3.3) serves as the foundation for *linear stability analysis*, a valuable technique for analyzing and understanding the stability of ODE solvers. The solution to this equation is given by $u(t) = e^{\lambda t}$, which grows rapidly if λ has a positive real part. Therefore, our primary interest lies in the case where $\Re(\lambda) < 0$, for which the analytical solution is stable, but our choice of solver may introduce *numerical instabilities*. When the FE method is applied to (3.3), we obtain the update formula

$$u_{n+1} = u_n + \Delta t \lambda u_n = u_n(1 + \Delta t \lambda),$$

and for the first step, with the initial condition $u(0) = 1$, we have

$$u_1 = 1 + \Delta t \lambda. \tag{3.4}$$

The analytical solution decays exponentially for $\Re(\lambda) < 0$, and it is natural to require that the numerical solution decreases monotonically. This leads to the requirement $|1 + \Delta t \lambda| \leq 1$. When λ is a negative real number, the time step must satisfy $\Delta t \leq -2/\lambda$ to ensure stability. It is important to note that meeting this stability criterion does not necessarily guarantee a highly accurate solution; the numerical solution may exhibit oscillate and differ substantially from the exact solution. Nevertheless, by selecting Δt to satisfy the stability criterion, we ensure that the solution, along with any spurious oscillations or other numerical artifacts, decays with time.

We have observed that the right-hand side of (3.4) contains critical information about the stability of the FE method. This expression is commonly referred to as the *stability function* or *amplification factor* of the method, and is often written as

$$R(z) = 1 + z.$$

For the FE method to be stable, all values of $\lambda \Delta t$ must satisfy $|R(\lambda \Delta t)| < 1$. This region of $\lambda \Delta t$ values in the complex plane is referred to as the method's *region of absolute stability* or its *stability region*. The stability region for the FE method is shown in the left panel of Figure 3.2, taking the form of a circle with center at $(-1, 1)$ and a radius of one. It is evident that if $\lambda \ll 0$, the requirement for $\lambda \Delta t$ to lie within this circle is quite restrictive for the choice of Δt.

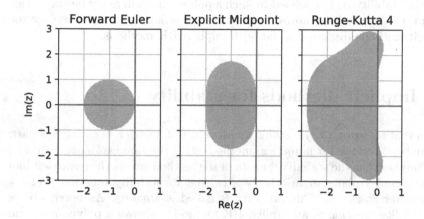

Fig. 3.2 Stability regions for explicit Runge-Kutta methods. From left: forward Euler, explicit midpoint, and the fourth order method given by (2.10)-(2.14).

We can easily extend the linear stability analysis to the other explicit RK methods introduced in Chapter 2. For instance, applying a single step of the explicit midpoint method given by (2.2)-(2.4) to (3.3) gives

$$u(\Delta t) = 1 + \lambda \Delta t + \frac{(\Delta t \lambda)^2}{2},$$

and we identify the stability function for this method as

$$R(z) = 1 + z + \frac{z^2}{2}.$$

The corresponding stability region is shown in the middle panel of Figure 3.2. For the fourth-order RK method defined in (2.10)-(2.14), the same steps yield the stability function

$$R(z) = 1 + z + \frac{z^2}{2} + \frac{z^3}{6} + \frac{z^4}{24},$$

and the stability region is shown in the right panel of Figure 3.2. We observe that the stability regions of these higher-order RK methods are slightly larger than that of the FE method. However, the difference is not very large, and when also considering the computational cost of each time step, the FE method is usually superior for problems where the time step is governed by stability.

It can be shown that the stability function for an s-stage explicit RK method is always a polynomial of degree $\leq s$, and it can easily be verified that the stability region defined by such a polynomial will never be very large. To obtain a significant improvement of this situation, we need to replace the explicit methods discussed so far with implicit RK methods.

3.2 Implicit methods for stability

Given that equation (3.3) is stable for all values of λ with a negative real part, it is natural to look for numerical methods with the same stability property. This implies that the stability domain of the method covers the entire left half of the complex plane, or in other words, that its stability function $|R(z)| \leq 1$ whenever $\Re(z) \leq 0$. This property is called *A-stability*. As noted above, the stability function of an explicit RK method is always a polynomial, and no polynomial can satisfy $|R(z)| < 1$ for all $z < 0$. Hence, there are no A-stable explicit RK methods. An even stronger stability requirement can be motivated by the fact that for $\lambda \ll 0$, the solution to (3.3) decays very rapidly. It is reasonable to expect the same behavior from the numerical solution, which leads to the requirement that $|R(z)| \to 0$ as $z \to -\infty$. This property is referred to as *stiff decay*, and an A-stable method that exhibits stiff decay is known as an *L-stable* method. For further detail, readers can refer to [1, 9].

The simplest implicit RK method is the backward Euler (BE) method, which can be derived in exactly the same way as the FE method, by approximating the derivative with a simple finite difference. The only difference from the FE method is that the right-hand side is evaluated at step $n+1$ rather than step n. For a general ODE, we have

$$\frac{u_{n+1} - u_n}{\Delta t} = f(t_{n+1}, u_{n+1}),$$

and if we rearrange the terms we get

$$u_{n+1} - \Delta t f(t_{n+1}, u_{n+1}) = u_n. \tag{3.5}$$

Although the derivation is similar to the FE method, there is a fundamental difference. In the BE method, the unknown variable u_{n+1} occurs as an argument in the right-hand side function $f(t, u)$. Therefore, for nonlinear f, (3.5) becomes a nonlinear algebraic equation that must be solved to determine

u_{n+1}, instead of having an explicit update formula as in the FE method. This requirement makes implicit methods more complex to implement than explicit methods, and they tend to require more computations per time step. However, as we will demonstrate later, the superior stability properties of implicit solvers still make them better suited for stiff problems.

We will consider the implementation of implicit solvers in Section 3.3 below, but let us first examine the stability of the BE method and other implicit RK solvers using the linear stability analysis introduced earlier. Applying the BE method to (3.3) yields

$$u_{n+1}(1 - \Delta t \lambda) = u_n,$$

and for the first time step, with $u(0) = 1$, we get

$$u_1 = \frac{1}{1 - \Delta t \lambda}.$$

The stability function of the BE method is therefore $R(z) = 1/(1-z)$, and its corresponding stability domain is shown in the left panel of Figure 3.3. The method is stable for all choices of $\lambda \Delta t$ *outside* the circle with a radius of one and centered at $(1,0)$ in the complex plane. This confirms that the BE method is a highly stable method. It is both A-stable, as its stability domain covers the entire left half of the complex plane, and L-stable, as the stability function satisfies $R(z) \to 0$ as $\Re(z) \to -\infty$.

The BE method fits into the general RK framework defined by (2.8)-(2.9) in Chapter 2, with a single stage ($s = 1$), and $a_{11} = b_1 = c_1 = 1$. Similar to the FE method discussed in Chapter 2, we can reformulate the method slightly to introduce a stage derivative and emphasize its connection to the RK family:

$$k_1 = f(t_n + \Delta t, u_n + \Delta t k_1), \tag{3.6}$$
$$u_{n+1} = u_n + \Delta t k_1. \tag{3.7}$$

The explicit midpoint and trapezoidal methods mentioned earlier also have their implicit counterparts. The implicit midpoint method is given by

$$k_1 = f(t_n + \Delta t/2, u_n + k_1 \Delta t/2), \tag{3.8}$$
$$u_{n+1} = u_n + \Delta t k_1, \tag{3.9}$$

while the implicit trapezoidal rule, or Crank-Nicolson method, is given by

$$k_1 = f(t_n, u_n), \tag{3.10}$$
$$k_2 = f(t_n + \Delta t, u_n + \Delta t k_2), \tag{3.11}$$
$$u_{n+1} = u_n + \frac{\Delta t}{2}(k_1 + k_2). \tag{3.12}$$

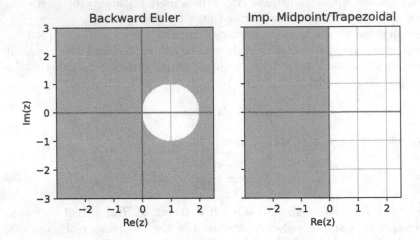

Fig. 3.3 Stability regions for the backward Euler method (left) and the implicit midpoint method and trapezoidal method (right).

Note that this formulation of the Crank-Nicolson is not very common, and it can be simplified by eliminating the stage derivatives and defining the method in terms of u_n and u_{n+1}. However, the given formulation in (3.10)-(3.12) highlights its implicit RK nature. The implicit nature of these methods is apparent from the formulas, as one of the stage derivatives must be found by solving an equation involving the nonlinear function f instead of using an explicit update formula. The Butcher tableaus of the three methods are given by

$$\begin{array}{c|c} 1 & 1 \\ \hline & 1 \end{array}, \quad \begin{array}{c|c} 1/2 & 1/2 \\ \hline & 1 \end{array}, \quad \begin{array}{c|cc} 0 & & \\ 1 & 0 & 1 \\ \hline & 1/2 & 1/2 \end{array}, \tag{3.13}$$

from left to right for backward Euler, implicit midpoint and the implicit trapezoidal method.

The implicit midpoint method and the implicit trapezoidal method share the same stability function, given by $R(z) = (2+z)/(2-z)$. The corresponding stability domain covers the entire left half-plane of the complex plane, as shown in the right panel of Figure 3.3. Both the implicit midpoint method and the trapezoidal method are therefore A-stable methods. However, since $R(z) \to 1$ as $z \to -\infty$, these methods lack stiff decay and are therefore not L-stable. In general, the stability functions of implicit RK methods are rational functions, i.e., given by

$$R(z) = \frac{P(z)}{Q(z)},$$

where P, Q are polynomials of degree at most s. This is in contrast to the stability functions of explicit methods, which are always polynomials of degree at most s, as mentioned in Section 3.1 above.

The accuracy of the implicit methods mentioned can be easily determined using a Taylor series expansion, as outlined in Section 1.5, to confirm that the backward Euler method is a first-order accurate method, while the implicit midpoint and trapezoidial methods are both second-order accurate. It is worth noting that while explicit Runge-Kutta methods with s stages have an order of accuracy $p \leq s$, implicit methods offer more flexibility in choosing the coefficients a_{ij}, potentially leading to higher accuracy for a given number of stages. In fact, the maximum order achievable for an implicit RK method is $p = 2s$, which is the case for the implicit midpoint method with $s = 1$ and $p = 2$. More advanced implicit RK methods will be explored later, but first, let us examine the implementation of the methods introduced so far.

3.3 Implementing Implicit Runge-Kutta Methods

In the previous section, we highlighted the superior stability and generally higher accuracy of the implicit methods. Following this discussion, one might question why IRK solvers are not the default choice for all ODE problems. The answer lies in the fact that they are implicit, so the stage derivatives are defined in terms of nonlinear equations rather than explicit formulae. This fact complicates the implementation of the methods and increases the computational cost of each time step. Due to this computational overhead, explicit solvers are usually more efficient for non-stiff problems, while implicit solvers are primarily suited for stiff ODEs.

For scalar ODEs, solving equations such as (3.5) or (3.8) using Newton's method is usually not overly challenging. However, when dealing with systems of ODEs, the task becomes more complex as we need to solve a system of coupled nonlinear equations. Applying Newton's method to a system of equations involves solving a system of linear equations at each iteration, which opens up a range of possible methods for solving these systems, as well as other solver choices and parameters that can be tuned to optimize the performance. In this text, our focus is primarily on understanding the fundamental concepts of IRK solvers, and we will not delve into the details of performance optimization. Therefore, we will base our implementation on built-in equation solvers from SciPy. We will start with the backward

Euler method, which is the simplest implicit method, but we will keep the implementation sufficiently general to be easily extendable to more advanced implicit methods. For a more detailed discussion on solver optimization and choices to enhance computational performance, interested readers can refer to references [1,9].

When examining the ODESolver class introduced in Chapter 2, we can observe that many administrative tasks involved in RK methods are the same for both implicit and explicit methods. Specifically, the initialization of solution arrays and the for-loop that advances the solution remain unchanged. However, advancing the solution from one step to the next differs significantly. Therefore, it is convenient to implement implicit solvers within the existing class hierarchy and let the ODESolver superclass handle the tasks of initializing the solver and the main solver loop. The different explicit methods introduced in Chapter 2 were realized through different implementations of the advance method. We can use the same approach for implicit methods, but since each step in implicit methods involves a few more operations it is useful to introduce a couple of additional methods. For instance, a concise implementation of the backward Euler method could appear as follows:

```python
from ODESolver import *
from scipy.optimize import root

class BackwardEuler(ODESolver):
    def stage_eq(self, k):
        u, f, n, t = self.u, self.f, self.n, self.t
        dt = self.dt
        return k - f(t[n] + dt, u[n] + dt * k)

    def solve_stage(self):
        u, f, n, t = self.u, self.f, self.n, self.t
        k0 = f(t[n], u[n])
        sol = root(self.stage_eq, k0)
        return sol.x

    def advance(self):
        u, f, n, t = self.u, self.f, self.n, self.t
        dt = self.dt
        k1 = self.solve_stage()
        return u[n] + dt * k1
```

Compared with the explicit solvers presented in Chapter 2, we have introduced two additional methods in our BackwardEuler class. The first method, stage_eq(self, k), is a Python implementation of (3.6), which defines the nonlinear equation for the stage derivative. The method takes the stage derivative k as input and returns the residual of (3.6). This formulation allows us to use SciPy's nonlinear equation solvers effectively. The actual solution of the stage derivative equation is handled in the solve_stage method. This method first computes an initial guess k0 for the stage derivative, and then passes this guess and the function stage_eq to SciPy's root function to solve the equation. The `root` function is a general tool for solving nonlinear equations of the form $g(x) = 0$, and we apply it to solve the stage equation $k_1 - f(t_n + \Delta t, u_n + \Delta t k_1) = 0$. The function returns an object of the `OptimizeResult` class, which includes the solution as an attribute x, along with numerous other attributes containing information about the solution process. For further details on the `OptimizeResult` and the `root` function, we refer to the SciPy documentation.

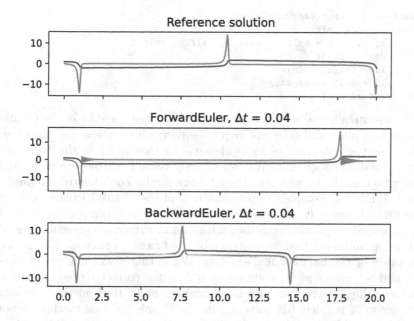

Fig. 3.4 Solutions of the Van der Pol model for $\mu = 10$, using the forward and backward Euler methods with $\Delta t = 0.04$.

We can demonstrate the superior stability of the BE method by revisiting the Van der Pol equation discussed earlier. Setting, for instance, $\mu = 10$, and solving the model using the FE and BE methods gives the plots shown in Figure 3.4. The top panel shows a reference solution computed with the SciPy `solve_ivp` solver using very low tolerance (`rtol=1e-10`). The middle panel shows the solution produced by FE with $\Delta t = 0.04$, showing visible oscillations in one of the solution components. Attempting to increase the time step further for this method leads to a divergent solution. On the other hand, the lower panel presents the solution obtained with BE, which is substantially more stable but still deviates from the reference solution in the top panel. With the BE method, increasing the time step further will still yield a stable solution, although it will deviate further from the exact solution. This simple experiment highlights the importance of considering both accuracy and stability when solving challenging systems of ODEs.[2]

Just as we did for the explicit methods in Chapter 2, it is possible to reuse code from the `BackwardEuler` class to implement other solvers. To facilitate extensive code reuse for a large range of implicit solvers, a slight modification of the code is required, which will be discussed in the next section. However, it is worth noting that a simple solver like the Crank-Nicolson method can be implemented with minimal changes to the `BackwardEuler` class. The class implementation could resemble the following:

```
class CrankNicolson(BackwardEuler):
    def advance(self):
        u, f, n, t = self.u, self.f, self.n, self.t
        dt = self.dt
        k1 = f(t[n], u[n])
        k2 = self.solve_stage()
        return u[n] + dt / 2 * (k1 + k2)
```

In this implementation, we leverage the fact that the stage k_1 in the Crank-Nicolson is explicit and does not require solving an equation. On the other hand, while the definition of k_2 is identical to that of k_1 in the backward Euler method. Consequently, We can directly reuse both the `stage_eq` and `solve_stage` methods, with only the `advance` method needing to be reimplemented. While this compact implementation of the Crank-Nicolson method allows for code reuse, it can be argued that it violates a common principle of object-oriented programming. Subclassing and inheritance represent an "is-a" relationship, implying that an instance of the `Crank-Nicolson` class is also an instance of the `BackwardEuler` class. While this works fine in the program, and is convenient for code reuse, it is not a correct representation of the relationship between the two numerical methods. Both methods belong to the group of implicit RK solvers, but the Crank-Nicolson method is not a special case of BackwardEuler. In the following sections, we will introduce an alternative class hierarchy that reflects this relationship and enables a compact implementation of RK methods using the general formulation in (2.8)-(2.9).

3.4 Implicit Methods of Higher Order

Similar to the ERK methods discussed in Chapter 2, the accuracy of IRK methods can be enhanced by increasing the number of stages. However, for implicit methods, we have more flexibility in selecting the parameters a_{ij},

---

[2]Note that the accompanying source code includes a script to generate Figure 3.4, as well as many other figures featured in the book. It is recommended to run these scripts independently and experiment with different time steps and parameters. Doing so will provide a better understanding of how the solvers work.

and this choice affects both the accuracy and computational complexity of the methods. In this section, we will explore two main branches of IRK methods: *fully implicit* methods and *diagonally implicit* methods. Both classes of methods are widely used and both have their advantages and drawbacks.

3.4.1 Fully Implicit RK Methods

The most general form of RK methods is known as fully implicit methods or FIRK methods. These solvers are defined by (2.8)-(2.9), with all coefficients a_{ij} (potentially) non-zero. When a method has more than one stage, this formulation implies that all stage derivatives depend on all other stage derivatives, so we need to determine them all at once by solving a single system of nonlinear equations. This operation is quite expensive, but the reward is that the FIRK methods have superior stability and accuracy for a given number of stages. A FIRK method with s stages can have order at most $2s$, which was the case for the implicit midpoint method in (3.8)-(3.9).

Many popular FIRK methods are based on combining standard numerical integration quadrature methods with the idea of *collocation*. Here, we present a brief overview of the derivation to illustrate the foundation shared by many important methods. For a comprehensive understanding, we recommend referring to references such as [9]. Recall from Chapter 2 that all RK methods can be viewed as approximations of equation (2.1), where the integral is approximated by a weighted sum. We set

$$u(t_{n+1}) = u(t_n) + \int_{t_n}^{t_{n+1}} f(t, u(t)) \approx u(t_n) + \sum_{i=1}^{s} b_i k_i, \qquad (3.14)$$

where b_i are the weights and k_i are the stage derivatives, which could be interpreted as approximations of the right-hand side function $f(t, u)$ at distinct time points $t_n + \Delta t c_i$.

Numerical integration is a well-established field in numerical analysis, and it is natural to choose the integration points c_i and weights b_i in (3.14) based on standard quadrature rules with known properties. Such quadrature rules are often derived by approximating the integrand with a polynomial which interpolates the function f at distinct points, and then integrating the polynomial exactly. A similar approach can be employed in deriving implicit RK methods. We approximate the solution u on the interval $t_n < t \leq t_{n+1}$ using a polynomial $P(t)$ of degree up to s, and require that $P(t)$ satisfies the ODE exactly at distinct points $t_n + c_i \Delta t$. This requirement, expressed as

$$P'(t_i) = f(t_i, P(t_i)), \quad t_i = t_n + c_i \Delta t, i = 1, \ldots, s. \qquad (3.15)$$

is known as collocation, and is a widely used concept in numerical analysis. It can be shown that, given a choice of quadrature points c_i, the collocation equations (3.15) uniquely determine the remaining coefficients a_{ij} and b_i of the method, see [1] for details.

A convenient approach to deriving FIRK methods is to choose a set of collocation points c_i, typically chosen from standard quadrature rules, and solve (3.15) to determine the remaining parameters. This strategy has led to families of FIRK methods based on common numerical integration rules. For instance, choosing c_i as Gauss points gives rise to the Gauss methods, which are the most accurate methods for a given number of stages, achieving order $2s$. The single-stage Gauss method corresponds to the implicit midpoint method introduced earlier, while the fourth-order Gauss method with $s = 2$ is defined by the Butcher tableau

$$
\begin{array}{c|cc}
\frac{3-\sqrt{3}}{6} & \frac{1}{4} & \frac{3-2\sqrt{3}}{12} \\
\frac{3+\sqrt{3}}{6} & \frac{3+2\sqrt{3}}{12} & \frac{1}{4} \\
\hline
 & \frac{1}{2} & \frac{1}{2}
\end{array}.
$$

The Gauss methods are A-stable but not L-stable. Since FIRK methods are primarily used for challenging stiff problems where stability is crucial, another family of FIRK methods, known as Radau IIA methods, is more commonly employed. These methods are based on Radau quadrature points, which include the right end of the integration interval (i.e., $c_s = 1$). The one-stage Radau IIA method is the backward Euler method, while the two- and three-stage versions are given by

$$
\begin{array}{c|cc}
1/3 & 5/12 & -1/12 \\
1 & 3/4 & 1/4 \\
\hline
 & 2/3 & 1/4
\end{array},
\qquad
\begin{array}{c|ccc}
\frac{4-\sqrt{6}}{10} & \frac{88-7\sqrt{6}}{360} & \frac{296-169\sqrt{6}}{1800} & \frac{-2+3\sqrt{6}}{225} \\
\frac{4+\sqrt{6}}{10} & \frac{296+169\sqrt{6}}{1800} & \frac{88+7\sqrt{6}}{360} & \frac{-2-3\sqrt{6}}{225} \\
1 & \frac{16-\sqrt{6}}{36} & \frac{16+\sqrt{6}}{36} & \frac{1}{9} \\
\hline
 & \frac{16-\sqrt{6}}{36} & \frac{16+\sqrt{6}}{36} & \frac{1}{9}
\end{array}.
$$

The Radau IIA methods exhibit order $2s - 1$, and their stability functions are $(s-1, s)$ Padé approximations of the exponential function, as described in [9]. For the two- and three-stage methods mentioned earlier, the stability functions are given by

$$
R(z) = \frac{1+z/3}{1-2z/3+z^2/6},
$$

$$
R(z) = \frac{1+2z/5+z^2/20}{1-3z/5+3z^2/20-z^2/60},
$$

respectively. The stability domains of these methods are depicted in Figure 3.5. Due to their L-stability, the Radau IIA methods are commonly used for solving stiff ODE systems. However, as noted above, the fact that all

$a_{ij} \neq 0$ complicates the implementation of the methods and makes each time step computationally expensive. All the s equations of (2.8) become fully coupled and need to be solved simultaneously. In the case of an ODE system comprising m equations, we must solve a system of ms nonlinear equations for each time step. We will come back the implementation of FIRK methods in Section 3.5, but let us first introduce a slightly simpler class of implicit RK solvers.

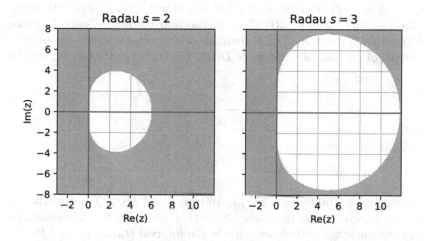

Fig. 3.5 The shaded area represents the stability region for two of the RadauIIA methods, with $s = 2$ (left) and $s = 3$ (right).

3.4.2 Diagonally Implicit RK Methods

Diagonally implicit RK (DIRK) methods, also known as semi-explicit methods, are a subclass of implicit RK methods. For DIRK methods, we have $a_{ij} = 0$ for all $j > i$. (Notice the small but important difference from the explicit methods, where we have $a_{ij} = 0$ for $j \geq i$.) The consequence of this choice is that the equation for a single stage derivative k_i does not involve stages k_{i+1}, k_{i+2}, and so on, and we can sequentially solve for the stage derivatives one by one. We still need to solve nonlinear equations to determine each k_i, but we can solve s systems of m equations rather than solving one large system to compute all stages simultaneously. This simplifies the implementation and reduces the computational cost per time step. However, the restriction on the method coefficients also reduces the accuracy and stability compared with FIRK methods. A general DIRK method with s stages has a maximum order of $s+1$, and methods optimized for stability typically have even lower order.

It is worth nothing that the implicit midpoint method discussed earlier technically falls under the category of DIRK methods. However, it is also a fully implicit Gauss method, and is not commonly referred to as a DIRK method. The distinction between FIRK and DIRK methods is meaningful only for $s > 1$. The Crank-Nicolson (implicit trapezoidal) method given by (3.10)-(3.12) is another example of a DIRK method, evident from the rightmost Butcher tableau in (3.13). These methods are, however, only A-stable, and it is possible to derive DIRK methods with better stability properties. An example of an L-stable, two-stage DIRK method of order two is given by

$$\begin{array}{c|cc} \gamma & \gamma & 0 \\ 1 & 1-\gamma & \gamma \\ \hline & 1-\gamma & \gamma \end{array}, \tag{3.16}$$

with stability function

$$R(z) = \frac{1 + z(1 - 2\gamma)}{(1 - z\gamma)^2}.$$

This method is A-stable for $\gamma > 1/4$, and for $\gamma = 1 \pm \sqrt{2}/2$ the method is L-stable and second-order accurate. Note that choosing $\gamma > 1$ means that we estimate the stage derivatives outside the interval (t_n, t_{n+1}), and for the last step beyond the time interval of interest. While this does not affect the stability or accuracy of the method, it may not be suitable for all ODE problems, and the most popular choice is therefore $\gamma = 1 - \sqrt{2}/2 \ (\approx 0.293)$. Notice also that in this method the two diagonal entries of a_{ij} are identical, with $a_{11} = a_{22} = \gamma$. This choice is very common in DIRK methods, and methods of this kind are known as *singly diagonally implicit* RK (SDIRK) methods. The main benefit of this structure is that the nonlinear equations for each stage derivative become very similar, which can be utilized when solving the equations using quasi-Newton methods. This benefit may not be very obvious for the examples in this book, since we rely on the generic `root` function from `scipy.optimize` to solve the nonlinear equations. However, if we wanted to improve the computational performance of the solvers, a natural place to start would be to implement a custom quasi-Newton solver that exploits the particular structure of the nonlinear equations. We will not go into the details of such an implementation here, but it is worth commenting on some aspects in order to understand why SDIRK methods are so widely used.

The central point is that when applying Newton's method to solve a general nonlinear system $g(u) = 0$, each iteration involves solving linear systems of the form $J_g \Delta u = -g(u^k)$, where Δu is the increment to the solution, u^k is the solution value at the previous iteration, and J_g is the Jacobian matrix of g, defined by

$$J_g = \frac{\partial g_i}{\partial u_j}.$$

For a general DIRK method, the nonlinear equation to compute stage derivative k_i is given by

$$k_i = f(t_n + c_i \Delta t, u_n + \Delta \sum_{j=1}^{i} a_{ij} k_j),$$

which can be written in the form $g(k_i) = 0$, with

$$g(k_i) = k_i - f\left(t_n + c_i \Delta t, u_n + \Delta \left(\sum_{j=1}^{i-1} a_{ij} k_j + a_{ii} k_i \right) \right).$$

Note that we have split the sum over the stage derivatives, highlighting that when solving for k_i, the values k_j for $j < i$ are already known. The Jacobian matrix J_g is found by differentiating g with respect to k_i, resulting in

$$J_g = I - \Delta t a_{ii} J_f,$$

where J_f is the Jacobian of the right-hand side function f. If we have $a_{ii} = \gamma$ for all stages, the Jacobian matrices J_g will also be identical, which can be exploited to optimize the solution of the linear systems. For more detailed information on solving nonlinear equations arising in SDIRK methods, refer to references such as [9].

Although we do not aim to present a complete overview of all subcategories of RK methods, one additional method class is worth mentioning. These are the so called ESDIRK (explicit singly diagonally implicit RK) methods, which are simply SDIRK methods where the first stage is explicit. The motivation behind these methods is that the nonlinear algebraic equations involved in the implicit methods are always solved with iterative methods, requiring an initial guess for the solution. For SDIRK methods, it is convenient to use the previous stage derivate as initial guess for the next one, which will usually provide a good initial guess. This approach is obviously not possible for the first stage, but an explicit formula for the first stage solves this problem. The simplest ESDIRK method is the implicit trapezoidal (Crank-Nicolson) method introduced above. A popular extension of this method is given by the following Butcher tableau:

$$\begin{array}{c|ccc} 0 & 0 & & \\ 2\gamma & \gamma & \gamma & 0 \\ 1 & \beta & \beta & \gamma \\ \hline & \beta & \beta & \gamma \end{array}' \tag{3.17}$$

with $\gamma = 1 - \sqrt{2}/2$ and $\beta = \sqrt{2}/4$. The resulting equations for each time step are

$$\begin{aligned} k_1 &= f(t_n, u_n), \\ k_2 &= f(t_n + 2\gamma\Delta t, u_n + \Delta t(\gamma k_1 + \gamma k_2)), \\ k_3 &= f(t_n + \Delta t, u_n + \Delta t(\beta k_1 + \beta k_2 + \gamma k_3)), \\ u_{n+1} &= u_n + \Delta t(\beta k_1 + \beta k_2 + \gamma k_3). \end{aligned}$$

This method can be interpreted as the sequential application of the trapezoidal method and a popular multistep solver called BDF2 (*backward differentiation formula* of order 2), and it is commonly known as the TR-BDF2 method. It is second order accurate, like the trapezoidal rule, but it is also L-stable, making it suitable for stiff problems.

3.5 Implementing Higher Order IRK Methods

In Section 3.3 we implemented two of the simplest implicit RK methods by a relatively small extension of the `ODEsolver` class hierarchy. We could easily continue this idea for the more complex IRK methods, and all the different methods could be realized by separate implementations of the three methods `solve_stage`, `stage_eq`, and `advance`. However, these three methods essentially implement the equations given by (2.8)-(2.9), which are common for all RK solvers. It is natural to look for an implementation that allows even more code reuse between the various methods, and we shall see that such a general implementation is indeed possible. However, it is still useful to treat the fully implicit methods and SDIRK methods separately, since the stage calculations of these two method classes are fundamentally different.

3.5.1 A Base Class for Fully Implicit Methods

One approach to implementing the fully implicit RK methods is to rewrite the `solve_stage`, `stage_eq`, and `advance` methods of the `BackwardEuler` class in a more general manner that can handle any number of stages and method parameters a_{ij}, b_i, and c_i. By adopting this approach, new methods can easily be implemented by specifying the number of stages and defining

the parameter values. In the methods we have discussed so far, the method
coefficients have been hard-coded within the mathematical expressions, often
inside the `advance` methods. However, with the generic approach, it is more
natural to define these coefficients as class attributes in the constructor. Following this general approach, a base class for implicit RK methods can be
defined as follows:

```python
from ODESolver import *
from scipy.optimize import root

class ImplicitRK(ODESolver):
    def solve_stages(self):
        u, f, n, t = self.u, self.f, self.n, self.t
        s = self.stages
        k0 = f(t[n], u[n])
        k0 = np.tile(k0,s)

        sol = root(self.stage_eq, k0)

        return np.split(sol.x, s)

    def stage_eq(self, k_all):
        a, c = self.a, self.c
        s, neq = self.stages, self.neq

        u, f, n, t = self.u, self.f, self.n, self.t
        dt = self.dt

        res = np.zeros_like(k_all)
        k = np.split(k_all, s)
        for i in range(s):
            fi = f(t[n] + c[i] * dt, u[n] + dt *
                    sum([a[i, j] * k[j] for j in range(s)]))
            res[i * neq:(1 + 1) * neq] = k[i] - fi

        return res

    def advance(self):
        b = self.b
        u, n, t = self.u, self.n, self.t
        dt = self.dt
        k = self.solve_stages()

        return u[n] + dt * sum(b_ * k_ for b_, k_ in zip(b, k))
```

Note that we assume that the method parameters are stored in NumPy arrays `self.a`, `self.b`, `self.c`, which need to be defined in subclasses. It
is important to note that, just as the `ODESolver` class discussed earlier, the
`ImplicitRK` class is intended as a pure base class for holding common code.
It is not meant to be used as a standalone solver class. In accordance with
the principles described in Section 2.2, we could make the abstract nature of
this class explicit by using the `abc` module, but for the present text we focus

on the fundamentals of the solvers and the class structure, keeping the code as simple and compact as possible.

The three methods of the `ImplicitRK` class are generalizations of the corresponding methods in the `BackwardEuler` class. They perform the same tasks but at a higher abstraction level and they rely on a bit of NumPy magic:

- The `solve_stages` method is a generalization of the `solve_stage` method above. Most of the lines are similar and should be self-explanatory. However, it is important to note that we are now implementing a general IRK method with s stages. Instead of solving a system of nonlinear equations for a single stage derivative, we solve a larger system to determine all s stage derivatives at once. The solution of this system is a one-dimensional array of length `self.stages * self.neq`, which contains all the stage derivatives. The line `k0 = np.tile(k0,s)` takes an initial guess `k0` for a single stage, and stacks it after itself s times to create the initial guess for all the stages, using NumPy's `tile` function.
- The `stage_eq` method is also a pure generalization of its `BackwardEuler` counterpart and performs the same tasks. The initial lines of this method are self-explanatory, while the `res = np.zeros_like(k_all)` creates an array of the appropriate length to hold the residual of the equation. For convenience, the line `k = np.split(k_all,s)` splits the array `k_all` into a list `k` that contains individual stage derivatives. This list is then used in the subsequent for loop on the next four lines. This loop, which forms the core of the method, implements equation (2.8), expressed as Python code and split over several lines for improved readability. The method returns the residual as a single array of length `self.stages * self.neq`, as expected by the SciPy `root` function.
- Finally, the `advance` method calls the `solve_stages` to compute all the stage derivatives, and then advances the solution using a general implementation of (2.9).

With the general base class in place, it becomes straightforward to implement new solvers by writing constructors that define the method coefficients. The following code implements the implicit midpoint and the two- and three-stage Radau methods:

```python
class ImplicitMidpoint(ImplicitRK):
    def __init__(self, f):
        super().__init__(f)
        self.stages = 1
        self.a = np.array([[1 / 2]])
        self.c = np.array([1 / 2])
        self.b = np.array([1])


class Radau2(ImplicitRK):
    def __init__(self, f):
        super().__init__(f)
        self.stages = 2
```

```
        self.a = np.array([[5 / 12, -1 / 12], [3 / 4, 1 / 4]])
        self.c = np.array([1 / 3, 1])
        self.b = np.array([3 / 4, 1 / 4])


class Radau3(ImplicitRK):
    def __init__(self, f):
        super().__init__(f)
        self.stages = 3
        sq6 = np.sqrt(6)
        self.a = np.array([[(88 - 7 * sq6) / 360,
                            (296 - 169 * sq6) / 1800,
                            (-2 + 3 * sq6) / (225)],
                           [(296 + 169 * sq6) / 1800,
                            (88 + 7 * sq6) / 360,
                            (-2 - 3 * sq6) / (225)],
                           [(16 - sq6) / 36, (16 + sq6) / 36, 1 / 9]])
        self.c = np.array([(4 - sq6) / 10, (4 + sq6) / 10, 1])
        self.b = np.array([(16 - sq6) / 36, (16 + sq6) / 36, 1 / 9])
```

Notice that we always define the method coefficients as NumPy arrays, even for the implicit midpoint method where they only contain a single number. This definition is necessary for the generic methods of the `ImplicitRK` class to work.

3.5.2 Base Classes for SDIRK and ESDIRK Methods

We could, in principle, implement both the SDIRK and ESDIRK methods in the same manner as the FIRK methods, by defining the method coefficients in the constructor and using the generic methods from the `ImplicitRK` base class. These generic methods would handle the cases where $a_{ij} = 0$ for $j > i$. However, the motivation behind the diagonally implicit methods is to avoid solving large systems of nonlinear equations, so it does not make much sense to implement them in this way. Instead, we should take advantage of the specific structure of the method coefficients and solve for the stage variables sequentially. This requires rewriting `solve_stages` and `stage_eq` methods from the base class. However, the `advance` method, which advances the solution to the next step, can remain unchanged as it is common to all RK methods.

Considering first the SDIRK methods, we can implement these as subclasses of the `ImplicitRK` class, which allows for some moderate code reuse and reflects the fact that SDIRK methods are special cases of implicit RK methods. To illustrate this implementation, let us first consider the two-stage SDIRK method defined by (3.16), and write out the equations for the stage derivatives to get a better view of the tasks involved. Inserting the coefficients from (3.16) into (2.8)-(2.9) gives

$$k_1 = f(t_n + \gamma \Delta t, u_n + \gamma \Delta t k_1), \tag{3.18}$$

$$k_2 = f(t_n + \Delta t, u_n + \Delta t((1 - \gamma)k_1 + \gamma k_2)), \tag{3.19}$$

$$u_{n+1} = u_n + \Delta t((1 - \gamma)k_1 + \gamma k_2). \tag{3.20}$$

Here, (3.18) is nearly identical to the equation defining the stage derivative in the backward Euler method, with the only difference being that Δt is replaced with $\gamma \Delta t$. Similarly, the only difference between (3.18) and (3.19) is the additional term $\Delta t(1 - \gamma)k_1$ inside the function call. In general, any stage equation for any DIRK method can be written as

$$k_i = f(t_n + c_i \Delta t, u_n + \Delta t(\sum_{j=0}^{i-1} a_{ij}k_j + \gamma k_i)), \tag{3.21}$$

where the sum inside the function call only includes previously computed stages.

Given the similarity of (3.21) with the stage equation from the backward Euler method, it is natural to implement the SDIRK stage equation as a generalization of the `stage_eq` method from the `BackwardEuler` class. To achieve this, we can create an SDIRK base class that contains the general versions of both the `stage_eq` and `solve_stages` methods. This base class can then be used as a foundation for deriving specific SDIRK solver classes. By writing the stage equations in this general form, it becomes straightforward to generalize the algorithm for looping through the stages and computing the individual stage derivatives. The complete base class implementation may appear as follows.

```
class SDIRK(ImplicitRK):
    def stage_eq(self, k, c_i, k_sum):
        u, f, n, t = self.u, self.f, self.n, self.t
        dt = self.dt
        gamma = self.gamma

        return k - f(t[n] + c_i * dt, u[n] + dt * (k_sum + gamma * k))

    def solve_stages(self):
        u, f, n, t = self.u, self.f, self.n, self.t
        a, c = self.a, self.c
        s = self.stages

        k = f(t[n], u[n])   # initial guess for first stage
        k_sum = np.zeros_like(k)
        k_all = []
        for i in range(s):
            k_sum = sum(a_ * k_ for a_, k_ in zip(a[i, :i], k_all))
            k = root(self.stage_eq, k, args=(c[i], k_sum)).x
            k_all.append(k)
        return k_all
```

The modified `stage_eq` method takes two additional parameters: the coefficient `c_i`, corresponding to the current stage, and the array `k_sum`, which holds the sum $\sum_{j=1}^{i-1} a_{ij}k_j$. These arguments need to be initialized correctly for each stage and passed as additional arguments to the SciPy `root` function. For convenience, we also assume that the method parameter γ has been stored as a separate class attribute. With the `stage_eq` method implemented in this general way, the `solve_stages` method simply needs to update the weighted sum of previous stages (`k_sum`), and pass this and the correct c value as additional arguments to the SciPy `root` function. The implementation uses a for loop to compute the stage derivatives sequentially and returns them as a list `k_all`.

As for the FIRK method classes, the only method we need to implement specifically for each solver class is the constructor, in which we define the number of stages and the method coefficients. A class implementation of the method in (3.16) may look as follows.

```
class SDIRK2(SDIRK):
    def __init__(self, f):
        super().__init__(f)
        self.stages = 2
        gamma = (2 - np.sqrt(2)) / 2
        self.gamma = gamma
        self.a = np.array([[gamma, 0],
                           [1 - gamma, gamma]])
        self.c = np.array([gamma, 1])
        self.b = np.array([1 - gamma, gamma])
```

Shifting our attention to the ESDIRK methods, they are identical to the SDIRK methods except for the first stage, and the potential for code reuse is obvious. The `stage_eq` method from the SDIRK base class can be directly reused in an ESDIRK solver class, since the equations to be solved for each stage are identical for SDIRK and ESDIRK solvers. However, the `solve_stages` method needs to be modified, since there is no need to solve a nonlinear equation for k1. Nevertheless, the modifications required are minimal since all stages $i > 1$ are identical. A possible implementation of the ESDIRK class can look as follows:

```
class ESDIRK(SDIRK):
    def solve_stages(self):
        u, f, n, t = self.u, self.f, self.n, self.t
        a, c = self.a, self.c
        s = self.stages
        k = f(t[n], u[n])  # initial guess for first stage
        k_sum = np.zeros_like(k)
        k_all = [k]
        for i in range(1, s):
            k_sum = sum(a_ * k_ for a_, k_ in zip(a[i, :i], k_all))
            k = root(self.stage_eq, k, args=(c[i], k_sum)).x
            k_all.append(k)
        return k_all
```

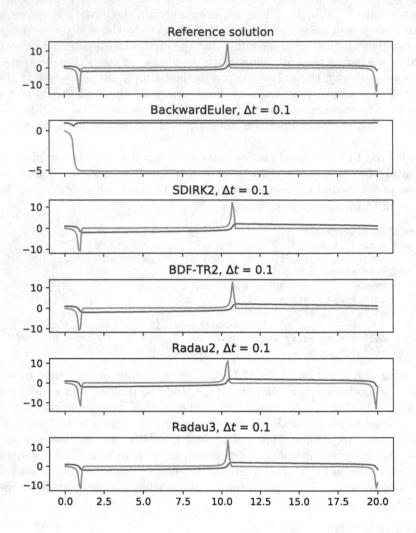

Fig. 3.6 Solutions of the Van der Pol model for $\mu = 10$ and $\Delta t = 0.1$, using implicit RK solvers of different accuracy.

Comparing with the SDIRK base class defined earlier, there are two small but important differences in the implementation of the solve_stages method. First, the result of the first function evaluation k = f(t[n],u[n]) is used directly as the first stage, by setting k_all = [k], instead of just serving as an initial guess for the nonlinear equation solver. Second, the for-loop for computing the remaining stages starts at i=1 rather than i=0.

With the ESDIRK base class at hand, individual ESDIRK methods can be implemented easily by defining the constructor, for instance:

```
class TR_BDF2(ESDIRK):
    def __init__(self, f):
        super().__init__(f)
        self.stages = 3
        gamma = 1 - np.sqrt(2) / 2
        beta = np.sqrt(2) / 4
        self.gamma = gamma
        self.a = np.array([[0, 0, 0],
                           [gamma, gamma, 0],
                           [beta, beta, gamma]])
        self.c = np.array([0, 2 * gamma, 1])
        self.b = np.array([beta, beta, gamma])
```

It should be noted that these class implementations have some potential weaknesses. One is that the solve_stages methods in the SDIRK and ES-DIRK classes are nearly identical, and most of the code is duplicated. Part of the purpose of implementing the solvers in a class hierarchy is to avoid code duplication, so this is clearly not optimal. However, avoiding duplicated code completely would require refactoring the classes a bit, to split the tasks performed in solve_stages into several methods. Since these tasks belong quite naturally together, splitting them up could make the code less readable and potentially less computationally efficient. Efficiency should always be a consideration when implementing numerical methods, although it is not a strong focus of this text. In the ESDIRK class, another choice that could be questioned is retaining the dimensions of the self.a coefficient array, and setting the entire first row to zero. Storing these zeros is unnecessary, and we could have omitted them and adjusted the for-loop in solve_stages accordingly. However, keeping the dimensions as they are helps maintain the link between the code and the mathematical formulation of RK methods.

Figure 3.6 illustrates the difference in accuracy between several IRK solvers. The chosen time step $\Delta t = 0.1$ is obviously too large for the backward Euler method, and the solution is not even close to the reference solution. The other solvers are the three-stage SDIRK method of order two, the two-stage Radau method of order three, and three-stage Radau method of order five. Further examples of SDIRK methods will be presented in Chapter 4, when we introduce RK methods with adaptive time step.

Chapter 4
Adaptive Time Step Methods

In practical computations, one seeks to achieve a desired accuracy with the minimum computational effort. For a given method, this requires finding the largest possible value of the time step Δt. In the previous chapters we kept the step size constant through the solution interval, but this is rarely the most efficient approach, since the error depends on the characteristics of the solution in addition to the step size. In smooth regions, larger time steps can be used without introducing significant error, while smaller time steps are needed in regions where the solution has rapid variations. This chapter extends the Runge-Kutta methods from the previous chapters to methods that select the time step automatically to control the error in the solution.

4.1 A Motivating Example

Many ODE models of dynamic systems have solutions that exhibit rapid variations in some intervals and remain nearly constant in others. A motivating example is a class of ODE models that describe the *action potential* of excitable cells, initially introduced by Hodgkin and Huxley [10]. These models play a crucial role in studying the electrophysiology of cells, including neurons and different types of muscle cells. The transmembrane potential, which is the difference in electrical potential between a cell's interior and its surroundings, is often the primary variable of interest. When an excitable cell, such as a neuron or muscle cell, undergoes electrical stimulation, it triggers a cascade of processes in the cell membrane, including to the opening and closing of various ion channels. The resulting flux of ions causes the membrane potential to transition from its resting negative state to approximately zero or slightly positive, before returning to its resting value. This process of *depolarization* followed by *repolarization* is called the action potential (see Figure 4.1). For a comprehensive overview of the Hodgkin-Huxley model and action potential models in general, refer to [11].

© The Author(s) 2024
J. Sundnes, *Solving Ordinary Differential Equations in Python*,
Simula SpringerBriefs on Computing 15,
https://doi.org/10.1007/978-3-031-46768-4_4

The potential value of adaptive time step methods becomes apparent when examining Figure 4.1. During the action potential, the solution changes rapidly whereas during periods of rest, it remains relatively constant over long time intervals. Similar behavior is observed in many types of ODE models and motivates the development of methods that can adjust the time step to match the solution's properties. These methods, commonly known as adaptive methods or methods with automatic time step control, are important components of all modern ODE software.

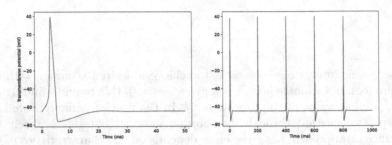

Fig. 4.1 Solution of the Hodgkin-Huxley model. The left panel shows a single action potential, while the right panel shows the result of stimulating the cell multiple times with a fixed period.

There are many possible approaches for automatically selecting the time step in numerical simulations. One intuitive strategy is to estimate the time step estimate based on the solution's dynamics, opting for a smaller time step during periods of rapid variations. This approach is commonly applied in adaptive solvers for partial differential equations (PDEs), where both the time step and space step can be chosen adaptively. It has also proven effective in specialized solvers for action potential models, as discussed in [15], where the time step is determined by the fluctuations in the transmembrane voltage. However, it is important to note that this method may not be universally applicable, and the criteria for choosing the time step must be carefully selected based on the characteristics of the problem at hand.

4.2 Choosing the Time Step Based on the Local Error

Adaptive time stepping methods aim to control the error in the solution, and it is natural to base the step selection on some form of error estimate. In Section 1.5 we computed the error at the end of the solution interval, and used it to confirm the theoretical convergence of the method. In principle, this global error could also be useful for selecting the time step, since we can simply redo the calculation with a smaller time step if the error is too large. However, for interesting ODE problems where the analytical solution is unavailable, this method of error estimation becomes complicated. Furthermore, the goal of adaptive time step methods is to dynamically select the time step as the solution progresses, to ensure that the final solution meets a specified error tolerance. This goal requires a different approach, which is based on estimating the local error for each step rather than relaying on the global error.

Assuming that we can estimate the local error for a given step, e_n, the goal is to choose the time step Δt_n so that the inequality

$$e_n < tol \tag{4.1}$$

is satisfied for all steps. The process of choosing Δt_n to ensure the satisfaction of (4.1) consists of two essential parts. First, we always check the inequality after performing a step. If it is satisfied, we accept the step and proceed with step $n+1$ as usual. If it is not satisfied, we reject the step and try again with a smaller Δt_n. The second part of the procedure involves choosing the next time step, Δt_{n+1}, if the current step was accepted, or a making a new guess for Δt_n if it was rejected. Interestingly, we will discover that the same formula, derived from our knowledge of the local error, can be applied in both cases.

For simplicity of notation, let us assume that step n was accepted with a time step Δt_n and a local error estimate $e_n < tol$. Our aim is now to choose Δt_{n+1} so that (4.1) is satisfied as sharply as possible, to avoid unnecessary computations. Hence, we aim to choose Δt_{n+1} such that $e_{n+1} \lessgtr tol$. Recall from 1.5 that for a method of global order p, the local error is of order $p+1$, so we have

$$e_n \approx C(\Delta t_n)^{p+1} \tag{4.2}$$
$$e_{n+1} \approx C(\Delta t_{n+1})^{p+1} \tag{4.3}$$

where we assume that the error constant C remains constant from one step to the next. Using (4.2), we can express C as

$$C = \frac{e_n}{(\Delta t_n)^{p+1}},$$

and by inserting this expression into (4.3), we obtain

$$e_{n+1} \approx \frac{e_n}{(\Delta t_n)^{p+1}}(\Delta t_{n+1})^{p+1}.$$

To achieve $e_{n+1} \approx tol$, we set

$$tol = e_{n+1} = \frac{e_n}{\Delta t_n^{p+1}} \Delta t_{n+1}^{p+1}$$

and rearrange to get the standard formula for time step selection

$$\Delta t_{n+1} = \left(\frac{tol}{e_n} \Delta t_n^{p+1} \right)^{1/(p+1)} .$$

We see that if $e_n \ll tol$, the formula will select a larger step size for the next step, while if $e_n \approx tol$ we get $\Delta t_{n+1} \approx \Delta t_n$. In practice, the formula is usually modified with a safety factor, i.e., we set

$$\Delta t_{n+1} = \eta \left(\frac{tol}{e_n} \Delta t_n^{p+1} \right)^{1/(p+1)} . \tag{4.4}$$

for some $\eta < 1$. The same formula can be used to choose a new step size Δt_n if the previous step was rejected, i.e., if $e_n > tol$.

   Although (4.4) provides a simple formula for the step size and works well for our example problems, more sophisticated methods have been derived. The task of choosing the time step to control the error is an optimal control problem, and successful methods based on control theory have been derived to control the error while avoiding abrupt changes in the step size. For detailed information and examples of such methods, refer to [9].

4.3 Estimating the Local Error

The inequality (4.1) and formula (4.4) provide the necessary tools to select the time step based on the local error e_n, and the remaining challenge is to develop a method for estimating this error. Since the analytical solution is unavailable, direct computation of the error is not feasible, but it can be estimated by comparing two numerical solutions of different accuracy. The general idea is to advance the solution from t_{n-1} to t_n using two methods of different accuracy, resulting in the regular solution, u_n, and a more accurate solution, $\hat{u}_n$. The difference $|\hat{u}_n - u_n|$ can then be used to estimate the local error for the solution u_n. The more accurate solution $\hat{u}_n$ can be computed in two ways: either by taking several "internal" time steps to advance from t_n to t_{n+1}, or by using a method with a higher order of accuracy. The former approach forms the basis of a technique referred to as *step doubling*, where the solution $\hat{u}_{n+1}$ is computed with the same method used for u_{n+1}, but with two steps of length $\Delta t/2$ instead of one step Δt. This naturally improves the accuracy of $\hat{u}_{n+1}$ compared with u_{n+1}, but the difference $|\hat{u}_{n+1} - u_{n+1}|$ is not large enough to be directly used as an error estimate. However, by combining this difference with the known order of the method, an error estimate can be derived. For further details, refer to [1]. The step doubling method is generally applicable and can provide a local error estimate for all ODE solvers. However, it is computationally expensive, and most modern ODE software relies on other techniques. The second approach for computing $\hat{u}_n$, to use a method with a higher order of accuracy, turns out to be particularly advantageous for RK methods. We shall see in the next section that it is possible to construct *embedded methods*, which provides an error estimate with very little additional computation.

4.3.1 Error Estimates from Embedded Methods

For a numerical method of order p, an estimate of the local error can be obtained by comparing the solution computed with a higher-order method (e.g., $p+1$), to the original solution. Since Δt is small, we have $\Delta t^{p+1} \ll \Delta t^p$, and we can directly estimate the error as $e_n = |u_n - \hat{u}_n|$. Computing these two solutions using two entirely different methods would be expensive. However, a more efficient approach is to use embedded methods, which are variations of a given RK method that achieve a different order of accuracy. Embedded methods use the same stage computations as the original method, making them relatively inexpensive to evaluate.

To introduce an embedded method for error estimation in the general RK method defined by (2.8)-(2.9), we define a separate set of weights $\hat{b}_i$, which advance the solution using the same k_i as the main method:

$$k_i = f(t_n + c_i \Delta t, y_n + \Delta t \sum_{j=1}^{s} a_{ij} k_j) \text{ for } i = 1, \ldots, s \qquad (4.5)$$

$$u_{n+1} = u_n + \Delta t \sum_{i=1}^{s} b_i k_i, \qquad (4.6)$$

$$\hat{u}_{n+1} = u_n + \Delta t \sum_{i=1}^{s} \hat{b}_i k_i. \qquad (4.7)$$

Although the main idea is to reuse the same stage computations to compute both $\hat{u}_{n+1}$ and u_{n+1}, it is not uncommon to introduce one additional stage in the method to obtain the error estimate. An RK method with an embedded method for error estimation is often referred to as an RK pair of order $n(m)$, where n is the order of the main method and m the order of the method used for error estimation. Butcher tableaus for RK pairs are written exactly as before, but with an extra line for the additional coefficients $\hat{b}$:

$$
\begin{array}{c|ccc}
c_i & a_{11} & \cdots & a_{1s} \\
\vdots & \vdots & & \vdots \\
c_s & a_{s1} & \cdots & a_{ss} \\
\hline
 & b_1 & \cdots & b_s \\
 & \hat{b}_1 & \cdots & \hat{b}_s
\end{array}
$$

As an example, let us consider the simplest possible embedded RK pair, which is obtained by combining Heun's method with the forward Euler method. The method is defined by the Butcher Tableau

$$
\begin{array}{c|cc}
0 & 0 & \\
1 & 1 & \\
\hline
 & 1 & 0 \\
\hline
 & 1/2 & 1/2
\end{array}
, \tag{4.8}
$$

which translates to the following formulas for advancing the two solutions:

$$
\begin{aligned}
k_1 &= f(t_n, u_n), \\
k_2 &= f(t_n + \Delta t, u_n + \Delta t k_1), \\
u_{n+1} &= u_n + \Delta t k_1, \\
\hat{u}_{n+1} &= u_n + \Delta t/2(k_1 + k_2).
\end{aligned}
$$

In the next section, we will see how this method pair can be implemented as an extension of the ODESolver hierarchy discussed earlier, before we introduce more advanced embedded RK methods in Section 4.5.

4.4 Implementing an Adaptive Solver

In the previous chapters we have successfully reused significant parts of the original ODESolver base class for a variety RK methods. The explicit RK methods only required reimplementing the advance method in subclasses, while the implicit methods needed a few additional methods and a minor redesign of the class structure. However, the solve method which contained the main solver loop, could be reused by all the subclasses. A closer inspection of this method reveals that the assumption of a fixed number of time steps is fundamental to the implementation, as it relies on a for-loop and fixed-size NumPy arrays. With the introduction of an adaptive step size, the number of steps is no longer fixed, necessitating significant changes to the solve method. In fact, the only part of the original ODESolver class that can be directly reused is the set_initial_condition method, providing only a modest benefit. Nevertheless, it still makes sense to implement the adaptive methods as subclasses of ODESolver, to benefit from this tiny code reuse and to highlight that an adaptive solver is a specialized case of a general ODE solver. Since most of the additional functionality needed by adaptive solvers is applicable to all adaptive methods, it makes sense to implement them in a generic base class. The following changes and additions are needed:

- A complete rewrite of the `solve` method, replacing the for-loop and NumPy arrays with lists and a while loop. Although lists are usually not preferred for computational tasks, their flexible size makes them suitable for adaptive time step methods. Additionally, it is natural to include more parameters in the `solve` function, allowing users to specify the tolerance, maximum step size and minimum step size.
- The `advance` method should be updated to return both the updated solution and the error estimate.
- The step selection formula in (4.4) must be implemented as a separate method.
- Adaptive methods usually include additional parameters, such as the safety factor η and the order p used in (4.4). These parameters can conveniently be defined as attributes in the constructor.

An implementation of the adaptive base class may look as follows:

```
from ODESolver import *
from math import isnan, isinf

class AdaptiveODESolver(ODESolver):
    def __init__(self, f, eta=0.9):
        super().__init__(f)
        self.eta = eta

    def new_step_size(self, dt, loc_error):
        eta = self.eta
        tol = self.tol
        p = self.order
        if isnan(loc_error) or isinf(loc_error):
            return self.min_dt

        new_dt = eta * (tol / loc_error)**(1 / (p + 1)) * dt
        new_dt = max(new_dt, self.min_dt)
        return min(new_dt, self.max_dt)

    def solve(self, t_span, tol=1e-3, max_dt=np.inf, min_dt=1e-5):
        """Compute solution for t_span[0] <= t <= t_span[1],
        using N steps."""
        t0, T = t_span
        self.tol = tol
        self.min_dt = min_dt
        self.max_dt = max_dt
        self.t = [t0]

        if self.neq == 1:
            self.u = [np.asarray(self.u0).reshape(1)]
```

```
    else:
        self.u = [self.u0]

    self.n = 0
    self.dt = 0.1 / np.linalg.norm(self.f(t0, self.u0))

    loc_t = t0
    while loc_t < T:
        u_new, loc_error = self.advance()
        if loc_error < tol or self.dt < self.min_dt:
            loc_t += self.dt
            self.t.append(loc_t)
            self.u.append(u_new)
            self.dt = self.new_step_size(self.dt, loc_error)
            self.dt = min(self.dt, T - loc_t, max_dt)
            self.n += 1
        else:
            self.dt = self.new_step_size(self.dt, loc_error)
    return np.array(self.t), np.array(self.u)
```

The constructor should be self-explanatory, but let us provide a few comments on the other two methods. The `new_step_size` method essentially implements (4.4) in Python, including tests to ensure that the selected step size falls within the user-defined range. We have also added a check to automatically set the new step size to the minimum step size if the computed error is infinity or not a number (`inf` or `nan`). This test is important for the solver's robustness because explicit methods will often diverge and return `inf` or `nan` values when applied to very stiff problems. By checking for these values and setting a low step size if they occur, we reduce the risk of complete solver failure. Although the computation may become inefficient with a small step size, it is preferable to unexpected failure.

The `solve` method has undergone significant changes compared to the `ODESolver` version. First, the parameter list has been expanded to include the tolerance, maximum time step and minimum time step. These values are stored as attributes and used in the main loop. The most notable changes start with the initialization of the `self.t` and `self.u` attributes, which are now lists of length one rather than fixed-size NumPy arrays. Note the somewhat cumbersome initialization of `self.u`, which includes an if-test to determine if we are solving a scalar ODE or a system. This initialization ensures that for scalar equations, `self.u[0]` is a one-dimensional array of length one, rather than a zero-dimensional array. The actual contents of these two data structures are the same, i.e., a single number, but they are treated differently by some NumPy tools, and it is useful to ensure that `self.u[0]`, `self.u[1]`, and so forth all have the same dimensions. The first step size is then calculated using a simplified version of the algorithm outlined in [8]. The for-loop has been replaced with a while-loop, since the number of steps is initially unknown. The call to the `advance`-method provides the updated solution and the estimated local error, after which we check if the local error is lower than the tolerance. If it is, the new time point and solution are appended to the corresponding lists, and the next time step is chosen based on the current step and the local error. The min and max operations ensure that the time step remains within the selected bounds and that the simulation ends at the final time T. If the constraint `loc_error < tol` is not satisfied, we simply compute a new time step and try again without updating the lists for the time and solution.

While the `solve` loop in the `AdaptiveODESolver` class is undoubtedly more complex than earlier versions, it is important to note that it still represents a simplifed version of an adaptive ODE solver. The aim here is to present the fundamental ideas and foster a general understanding of how these solvers are implemented. Consequently, we have included only the most essential components, and certain limitations and simplifications should be acknowledged:

- The step size selection formula in (4.4), implemented in the method `new_step_size`, could be replaced with more sophisticated algorithms. For more details, refer to sources such as [3, 9].
- The formula for selecting the initial step is quite basic and primarily aims to prevent extremely poor initial choices. More advanced algorithms have been developed, and for additional information, consult references like [8,9] for details.
- The initial `if`-test within the solver loop is not the most robust, since it will proceed and move forward if the minimum step size is reached, even if the error is excessively large. A robust solver should provide a warning to the user in such cases where the requested tolerance cannot be achieved.

Despite these and other limitations, the adaptive solver class works as intended and captures the essential behavior of adaptive ODE solvers.

With the `AdaptiveODESolver` base class available, specific solvers can be implemented by creating tailored versions of the `advance` method and the constructor. The order of the method is used in the time step selection and therefore needs to be defined as an attribute. For example, an implementation of the Euler-Heun method pair mentioned earlier could appear as follows:

```python
class EulerHeun(AdaptiveODESolver):
    def __init__(self, f, eta=0.9):
        super().__init__(f, eta)
        self.order = 1

    def advance(self):
        u, f, t = self.u, self.f, self.t
        dt = self.dt
        k1 = f(t[-1], u[-1])
        k2 = f(t[-1] + dt, u[-1] + dt * k1)
        high = dt / 2 * (k1 + k2)
        low = dt * k1
        unew = u[-1] + low
        error = np.linalg.norm(high - low)
        return unew, error
```

After calculating the derivatives k1 and k2 for the two stages, the method proceeds to compute the updates for both the high and low order solutions. The low order solution is used to advance the overall solution, while the difference between the high and low order solutions serves as the error estimate. The method then returns the updated solution and the error, which are needed by the solve method implemented in the base class described earlier.

Since we have two methods with different levels of accuracy, one might wonder whether it would be better to advance the solution using the more accurate method rather than the less accurate one. This choice would certainly yield a reduced local error, but the drawback is that we would no longer have a proper error estimate for the method used to integrate the solution. We can use the more accurate solution to estimate the error of the less accurate, but not the other way around. Nevertheless, this approach, known as *local extrapolation* [8], is still used by many popular RK pairs, as we will observe in the examples below. Even though the error estimate may not be precise for the method used to integrate the solution, it still works well as a tool for selecting the time step. In the implementation above, it is straightforward to experiment with this choice by replacing low with high when assigning the value to unew. By doing so, we can observe the impact on the error and the number of time steps.

4.5 More Advanced Embedded RK Methods

There are numerous examples of explicit RK pairs of higher order than the 1(2) pair defined by (4.8). We will not provide an exhaustive list here, but mention two particularly popular methods that have been implemented in various software packages. The first method, known as the Fehlberg 4(5) or RKF45 method [5] is defined by the following Butcher tableau:

$$
\begin{array}{c|cccccc}
0 & & & & & & \\
\frac{1}{4} & \frac{1}{4} & & & & & \\
\frac{3}{8} & \frac{3}{32} & \frac{9}{32} & & & & \\
\frac{12}{13} & \frac{1932}{2197} & -\frac{7200}{2197} & \frac{7296}{2197} & & & \\
1 & \frac{439}{216} & -8 & \frac{3680}{513} & -\frac{845}{4104} & & \\
\frac{1}{2} & -\frac{8}{27} & 2 & -\frac{3544}{2565} & \frac{1859}{4104} & -\frac{11}{40} & \\
\hline
& \frac{25}{216} & 0 & \frac{1408}{2565} & \frac{2197}{4104} & -\frac{1}{5} & 0 \\
& \frac{16}{135} & 0 & \frac{6656}{12825} & \frac{28561}{56430} & -\frac{9}{50} & \frac{2}{55}
\end{array}
\tag{4.9}
$$

In this tableau, the coefficients in the first line (b_i) correspond to a fourth-order method, while the coefficients in the last line $(\hat{b}_i)$ correspond to a fifth-order method. The implementation of the RKF45 method is similar to the

Euler-Heun pair, but due to the increased number of stages and coefficients, the `advance` method becomes more complex:

```python
class RKF45(AdaptiveODESolver):
    def __init__(self, f, eta=0.9):
        super().__init__(f, eta)
        self.order = 4

    def advance(self):
        u, f, t = self.u, self.f, self.t
        dt = self.dt
        c2 = 1/4; a21 = 1/4;
        c3 = 3/8; a31 = 3/32; a32 = 9/32
        c4 = 12/13; a41 = 1932/2197; a42 = -7200/2197; a43 = 7296/2197
        c5 = 1; a51 = 439/216; a52 = -8; a53 = 3680/513;
        a54 = -845/4104
        c6 = 1/2; a61 = -8/27; a62 = 2; a63 = -3544/2565;
        a64 = 1859/4104; a65 = -11/40
        b1 = 25/216; b2 = 0; b3 = 1408/2565; b4 = 2197/4104;
        b5 = -1/5; b6 = 0
        bh1 = 16/135; bh2 = 0; bh3 = 6656/12825; bh4 = 28561/56430;
        bh5 = -9/50; bh6 = 2/55

        k1 = f(t[-1], u[-1])
        k2 = f(t[-1] + c2 * dt, u[-1] + dt * (a21 * k1))
        k3 = f(t[-1] + c3 * dt, u[-1] + dt * (a31 * k1 + a32 * k2))
        k4 = f(t[-1] + c4 * dt, u[-1] + dt *
               (a41 * k1 + a42 * k2 + a43 * k3))
        k5 = f(t[-1] + c5 * dt, u[-1] + dt *
               (a51 * k1 + a52 * k2 + a53 * k3 + a54 * k4))
        k6 = f(t[-1] + c6 * dt, u[-1] +
               dt * (a61 * k1 + a62 * k2 + a63 * k3
                     + a64 * k4 + a65 * k5))

        low = dt * (b1 * k1 + b3 * k3 + b4 * k4 + b5 * k5)
        high = dt * (bh1 * k1 + bh3 * k3 + bh4 * k4
                     + bh5 * k5 + bh6 * k6)
        unew = u[-1] + low
        error = np.linalg.norm(high - low)

        return unew, error

        return unew, error
```

The `advance` method could be written more concisely, but we have chosen to maintain the structure of the explicit RK methods introduced earlier.

Another well-known and widely used pair of ERK methods is the Dormand-Prince method [4], which is a seven-stage method with the following coefficients:

$$
\begin{array}{c|ccccccc}
0 \\
\frac{1}{5} & \frac{1}{5} \\
\frac{3}{10} & \frac{3}{40} & \frac{9}{40} \\
\frac{4}{5} & \frac{44}{45} & -\frac{56}{15} & \frac{32}{9} \\
\frac{8}{9} & \frac{19372}{6561} & -\frac{25360}{2187} & \frac{64448}{6561} & -\frac{212}{729} \\
1 & \frac{9017}{3168} & -\frac{355}{33} & \frac{46732}{5247} & \frac{49}{176} & -\frac{5103}{18656} \\
1 & \frac{35}{84} & 0 & \frac{500}{1113} & \frac{125}{192} & -\frac{2187}{6784} & \frac{11}{84} \\
\hline
y_n & \frac{35}{384} & 0 & \frac{500}{1113} & \frac{125}{192} & -\frac{2187}{6784} & \frac{11}{84} & 0 \\
\hat{y}_n & \frac{5179}{57600} & 0 & \frac{7571}{16695} & \frac{393}{640} & -\frac{92097}{339200} & \frac{187}{2100} & \frac{1}{40}
\end{array}
$$

This method has been optimized for the local extrapolation approach mentioned above, where the higher order method is used for advancing the solution and the less accurate method is used for step size selection. The implementation is otherwise similar to the RKF45 method. The Dormand-Prince method has been implemented in many software tools, including the popular **ode45** function in Matlab (The Math Works, Inc. MATLAB. Version 2023a).

Implicit RK methods can also incorporate embedded methods. The underlying idea is the same as for explicit methods, although step size selection tends to be more challenging for stiff problems. A crucial requirement for stiff problems is that both the main method and the error estimator must have good stability properties. Stiff problems pose challenges for error control algorithms, and simple algorithms such as (4.4) often experience large fluctuations in step size and local error. For a detailed discussion of these challenges, refer to [1, 9].

As an example of an implicit method with error control, we can extend the TR-BDF2 method in (3.17) to include a third order method for error estimation. The extended Butcher tableau is

$$
\begin{array}{c|ccc}
0 & 0 \\
2\gamma & \gamma & \gamma & 0 \\
1 & \beta & \beta & \gamma \\
\hline
 & \beta & \beta & \gamma \\
 & \frac{1-\beta}{3} & \frac{3\beta+1}{3} & \frac{\gamma}{3}
\end{array}
, \tag{4.10}
$$

where $\gamma = 1 - \sqrt{2}/2$, $\beta = \sqrt{2}/4$, and the bottom line of coefficients defines the third-order method. This third-order method is not L-stable, so for stiff problems it is preferable to advance the solution using the second-order method and use the more accurate one for time step control. Achieving L-stability for both methods of an embedded RK pair is ideal but often impossible, and we need to accept somewhat weaker stability requirements for the error estimator, as discussed in [13].

When implementing the adaptive TR-BDF2 and other implicit methods, we need to combine features from the **AdaptiveODESolver** class mentioned earlier with the tools from the **ImplicitRK** hierarchy introduced in Chapter 3. Specifically, an adaptive implicit RK method requires the `solve` and `new_step_size` methods from `AdaptiveODESolver`, while all the code for computing the stage derivatives can be reused directly from the `ImplicitRK` classes. A convenient approach to reuse functionality from two different classes is to use *multiple inheritance*, where we define a new class as a subclass of two different base classes. For instance, a base class for adaptive ESDIRK methods may be implemented as follows:

```
class AdaptiveESDIRK(AdaptiveODESolver,ESDIRK):
```

This simply states that the new class inherits all the methods from both the `AdaptiveODESolver` class and the `ImplicitRK` class. The general design of the `ImplicitRK` class mentioned earlier was to define the method coefficients in the constructor and use a generic `advance` method, making it convenient to use the same method for adaptive implicit methods. However, the `advance` method needs to be overridden in our `AdaptiveImplicitRK` base class from `ImplicitRK` as we need the method to return the error in addition to the updated solution. All other methods can be reused directly from either `AdaptiveODESolver` or `ImplicitRK`. Therefore, a suitable implementation of the new class may look like:

```
class AdaptiveESDIRK(AdaptiveODESolver, ESDIRK):
    def advance(self):
        b = self.b
        e = self.e
        u = self.u
        dt = self.dt
        k = self.solve_stages()
        u_step = dt * sum(b_ * k_ for b_, k_ in zip(b, k))
        error = dt * sum(e_ * k_ for e_, k_ in zip(e, k))

        u_new = u[-1] + u_step
        error_norm = np.linalg.norm(error)
        return u_new, error_norm
```

In this implementation, we assume that the constructor defines all the RK method parameters used earlier, as well as a set of parameters `self.e`, defined by $e_i = b_i - \hat{b}_i$, for $i = 1,\ldots,n$, which are used in error calculations. Except for the two lines computing the error, the method is identical to the generic `advance` method from the `ImplicitRK` class used by all the previous subclasses. Consequently, one might wonder whether this method should have been placed in a general base class for implicit RK methods, such as `AdaptiveImplicitRK`, so that it could be used in adaptive versions of the SDIRK, ESDIRK, and Radau classes. However, adaptive versions of the Radau methods use a slightly different calculation of the error, since it is not possible to construct an embedded method of order $p-1$ for a Radau method of order p. Thus, for the adaptive solvers, the advance method is slightly less general, and it is more convenient to implement it separately for the ESDIRK methods. Further details on adaptive versions of the Radau methods may be found in [9].

Although multiple inheritance provides a convenient way to reuse the functionality of our existing classes, it comes with the risk of somewhat complex and confusing class hierarchies. In particular, the fact that our AdaptiveESDIRK class inherits from AdaptiveODESolver and ESDIRK, which are both subclasses of ODESolver, may give rise to a well-known ambiguity referred to as the *diamond problem*. The problem would arise if, for instance, we were to define a method in ODESolver, override it with special versions in both AdaptiveODESolver and ESDIRK, and then call it from an instance of AdaptiveESDIRK. Would we then call the version implemented in AdaptiveODESolver or the one in ESDIRK? The answer is determined by Python's so-called *method resolution order* (MRO), which decides which method to inherit first based on its "closeness" in the class hierarchy and then on the order of the base classes in the class definition. In our particular example the AdaptiveESDIRK class is equally close to AdaptiveODESolver and ESDIRK, since it is a direct subclass of both. The method called would therefore be the version from AdaptiveODESolver, since this is listed first in the class definition. In our relatively simple class hierarchy there are no such ambiguities, and even if we use multiple inheritance it should not be too challenging to determine which methods are called, but it is a potential source of confusion that is worth being aware of.

Now that we have the AdaptiveESDIRK base class available, we can implement an adaptive version of the TR-BDF2 method as follows:

```python
class TR_BDF2_Adaptive(AdaptiveESDIRK):
    def __init__(self, f, eta=0.9):
        super().__init__(f, eta)  # calls AdaptiveODESolver.__init__
        self.stages = 3
        self.order = 2
        gamma = 1 - np.sqrt(2) / 2
        beta = np.sqrt(2) / 4
        self.gamma = gamma
        self.a = np.array([[0, 0, 0],
                           [gamma, gamma, 0],
                           [beta, beta, gamma]])
        self.c = np.array([0, 2 * gamma, 1])
        self.b = np.array([beta, beta, gamma])
        bh = np.array([(1 - beta) / 3, (3 * beta + 1) / 3, gamma / 3])
        self.e = self.b - bh
```

To illustrate the use of this solver class, we may return to the Hodgkin-Huxley model introduced earlier in this chapter. Assuming the model is implemented as a class in a file `hodgkinhuxley.py`, the following code solves the model and plots the transmembrane potential:

```python
from AdaptiveImplicitRK import TR_BDF2_Adaptive
from hodgkinhuxley import HodgkinHuxley
import matplotlib.pyplot as plt

model = HodgkinHuxley()
u0 = [-45, 0.31, 0.05, 0.59]
t_span = (0, 50)
tol = 0.01

solver = TR_BDF2_Adaptive(model)
solver.set_initial_condition(u0)

t, u = solver.solve(t_span, tol)

plt.plot(t, u[:, 0])
plt.show()
```

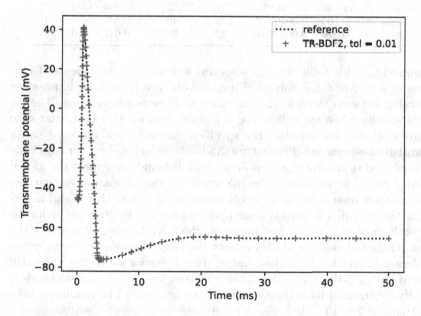

Fig. 4.2 Solution of the Hodgkin-Huxley model. The solid line is a reference solution computed with SciPy `solve_ivp`, while the +-marks are the time steps chosen by the adaptive TR-BDF2 solver.

A plot of the solution is shown in Figure 4.2, where the +-marks represent the time steps chosen by the adaptive TR-BDF2 solver. It is apparent that larger time steps are used in quiescent regions while smaller steps are employed in regions with rapid solution variations. A more quantitative view of the solver behavior, for three different solvers, is shown in the table below. Each method was applied with three different tolerance values over a time interval from 0 to 50ms, using default choices for the maximum and minimum time steps. The "Error" column provides an estimate of the global error, calculated based on a reference solution obtained using SciPy's `solve_ivp` function. The "Steps" column indicates the number of accepted time steps, while "Rejected" indicates the total number of rejected steps. The last two columns display the minimum and maximum time steps observed during the computation.

Solver	Tolerance	Error	Steps	Rejected	Δt_{max}	Δt_{min}
TR-BDF2	1.000	0.0336961	24	9	10.533	0.00791
TR-BDF2	0.100	0.0175664	43	14	9.705	0.00791
TR-BDF2	0.010	0.0028838	83	22	5.328	0.00791
RKF45	1.000	0.6702536	192	113	2.204	$1.0 \cdot 10^{-5}$
RKF45	0.100	0.0934201	118	58	1.093	$1.0 \cdot 10^{-5}$
RKF45	0.010	0.0054336	123	34	1.297	0.00791
EulerHeun	1.000	0.7790353	158	35	1.849	0.00791
EulerHeun	0.100	0.0016577	220	40	0.836	0.00791
EulerHeun	0.010	0.0014654	432	36	0.918	0.00251

The numbers in this table illustrate several well-known properties and limitations of adaptive ODE solvers. First, we observe that there is no close relationship between the selected tolerance and the resulting error. The error gets smaller when we reduce the tolerance, and for this particular case the error is always smaller than the specified tolerance, but the error varies substantially between the different methods. As mentioned earlier, the time step is selected to control the *local error*, and although we expect the global error to decrease as we reduce the tolerance, we cannot guarantee that the global error will be smaller than the tolerance. Second, the RKF45 and Euler-Heun methods exhibit relatively poor performance and inconsistent behavior as the tolerance is reduced. For instance, the RKF45 method requires the highest number of steps, and also rejects the largest number of steps, when the tolerance is set to the highest value. This behavior stems from the stiff nature of Hodgkin-Huxley model, where the time step for explicit methods is primarily determined by stability rather than accuracy. The minimum time step Δt_{min} of $1.0 \cdot 10^{-5}$ is a result of divergence issues that automatically set the time step to the specified lower bound. In most of the other combinations of method and tolerance, the smallest observed time step is the first one, selected by the simple formula within the `solve` method. There is room for improvement in this area, and the overall performance of RKF45 for stiff problems could be improved with a more sophisticated step size controller. However, it is important to note that for stiff problems, explicit solvers will never achieve the same level of performance as implicit solvers.

The ideas and tools introduced in this chapter are fundamental to all RK methods with error control and automatic time step selection. These ideas are fairly simple, and, as illustrated in Figure 4.2, give rise to methods that effectively adapt the time step to control the error. However, there are many practical considerations in implementing these methods, and we have only scratched the surface. For example, the time step control formula in (4.4) could be refined using more sophisticated models derived from control theory. [7] The initial time step selection, as indicated by the smallest step Δt_{min} being the first one for most solvers in the table, could also be improved. Furthermore, adjusted error estimates tailored for stiff systems have been proposed [9]. For a comprehensive discussion and detailed exploration of automatic time step control, we recommend referring to [1] and [8, 9].

Chapter 5
Modeling Infectious Diseases

Throughout this book we have focused entirely on *solving* ODEs, without delving deeply into their origins or applications. In the present chapter we shift our attention to *modeling* with ODEs, by exploring a widely studied class of ODE models that describe the spread of infectious diseases. This group of models serves as a good example of how a complex phenomenon can be modeled using relatively simple systems of ODEs. We will derive these models based on a set of fundamental assumptions and discuss the limitations that arise from these assumptions. Although we consider a single application and a single class of models, the fundamental steps of the modeling process are applicable to a wide range of real-world phenomena.

5.1 Derivation of the SIR model

Our objective is to develop a model that captures the dynamics of infectious disease transmission within a population. This subject is of great scientific and societal interest and has been studied by scientists for centuries, gaining even more attention in recent times for obvious reasons. In the early 1900s, Kermack and McKendrick [12] introduced the classical model for predicting epidemic dynamics, known as the SIR model. The name 'SIR' derives from the three categories it describes: Susceptible, Infected, and Recovered (or, alternatively, Removed). However, it is important to note that the spread of disease in a population is a very complex process, and to construct an ODE-based model we need to make a number of simplifying assumptions. The most important assumption is that we do not consider individuals as discrete entities but focus solely on the total population and the flow of people among the three aforementioned categories. We assume a perfectly mixed population confined to a specific area, disregarding spatial movement of the disease, and focusing solely on its temporal evolution. The first model we will derive is very simple, but it can be easily extended to encompass models used worldwide

J. Sundnes, *Solving Ordinary Differential Equations in Python*,
Simula SpringerBriefs on Computing 15,
https://doi.org/10.1007/978-3-031-46768-4_5

by health authorities for predicting the spread of diseases such as Covid-19, flu, Ebola, HIV, and others.

In the first version of the model we track the three categories of people mentioned above:

- S: susceptibles - who can get the disease
- I: infected - who have developed the disease and can infect susceptibles
- R: recovered - who have recovered and gained immunity

Mathematically, we represent these categories as functions $S(t)$, $I(t)$, $R(t)$, which denote the number of people in each category. Our goal is now to derive a set of equations for $S(t)$, $I(t)$, $R(t)$, and then solve these equations to predict the spread of the disease.

To derive the model equations, we first consider the dynamics over a time interval Δt, and our goal is to derive mathematical expressions for how many people that transition between the three categories during this time interval. The crucial aspect of the model lies in describing how individuals transition from S to I, i.e., how susceptible individuals become infected by those already infected. Since infectious diseases are primarily transmitted through direct interactions, we need to establish mathematical descriptions of the number of interactions between susceptible and infected individuals. We make the following assumptions:

- An individual in the S category interacts with an approximately constant number of people each day, making the number of interactions in a time interval Δt proportional to Δt.
- The probability of one of these interactions involving an infected person is proportional to the ratio of infected individuals to the total population, i.e., to I/N, with $N = S + I + R$.

Based on these assumptions, the probability of a single susceptible person becoming infected is proportional to $\Delta t I/N$. The total number of infections can be expressed as $\beta S I/N$, where β is a constant representing the probability of an infected person encountering and infecting a susceptible person. The value of β depends on the infectiousness of the disease and the behavior of the population, as further discussed below. Infections of new individuals lead to a decrease in S and a corresponding gain in I, resulting in the following equations:

$$S(t+\Delta t) = S(t) - \Delta t \beta \frac{S(t)I(t)}{N},$$

$$I(t+\Delta t) = I(t) + \Delta t \beta \frac{S(t)I(t)}{N}.$$

These two equations represent the key component of all the models covered in this chapter. They are formulated as *difference equations*, and they can easily be transformed to ODEs. More advanced models are typically derived by adding more categories and more transitions between them, but the individual transitions are very similar to those presented here.

Fig. 5.1 Graphical representation of the simplest SIR-model, where people move from being susceptible (S) to being infected (I) and then reach the recovered (R) category with immunity against the disease.

We also need to model the transition of people from the I to the R category. Again considering a small time interval Δt, it is reasonable to assume that a fraction $\Delta t\,\nu$ of the infected individuals recover and move to the R category. Here ν is a constant that describes the time dynamics of the disease. The increase in R is given by:

$$R(t+\Delta t) = R(t) + \Delta t\,\nu I(t),$$

Additionally, we need to subtract the same term in the balance equation for I, since individuals move from I to R. Therefore, we have:

$$I(t+\Delta t) = I(t) + \Delta t\,\beta S(t)I(t) - \Delta t\,\nu I(t).$$

We now have three equations for S, I, and R:

$$S(t+\Delta t) = S(t) - \Delta t\,\beta\frac{S(t)I(t)}{N}, \tag{5.1}$$

$$I(t+\Delta t) = I(t) + \Delta t\,\beta\frac{S(t)I(t)}{N} - \Delta t\,\nu I(t), \tag{5.2}$$

$$R(t+\Delta t) = R(t) + \Delta t\,\nu I(t). \tag{5.3}$$

These equations form a system of difference equations, as discussed in more detail in Appendix A. Although we could solve the equations in their current form using techniques from Appendix A, it is more convenient to formulate the model as a system of ODEs and apply the ODE solvers derived in previous chapters.

To turn the difference equations into ODEs, we first divide all equations by Δt and rearrange, to get

$$\frac{S(t+\Delta t) - S(t)}{\Delta t} = -\beta \frac{S(t)I(t)}{N}, \tag{5.4}$$

$$\frac{I(t+\Delta t) - I(t)}{\Delta t} = \beta t \frac{S(t)I(t)}{N} - \nu I(t), \tag{5.5}$$

$$\frac{R(t+\Delta t) - R(t)}{\Delta t} = \nu I(t). \tag{5.6}$$

We see that by letting $\Delta t \to 0$, we get derivatives on the left-hand side:

$$S'(t) = -\beta \frac{SI}{N}, \tag{5.7}$$

$$I'(t) = \beta \frac{SI}{N} - \nu I, \tag{5.8}$$

$$R'(t) = \nu I, \tag{5.9}$$

where, as above, $N = S + I + R$.[1] If we add the three equations, we see that $N'(t) = S'(t) + I'(t) + R'(t) = 0$, which means the total population N is constant. The equations (5.7)-(5.9) form a system of three ODEs, which we will solve to find the unknown functions $S(t)$, $I(t)$, $R(t)$. To solve these equations, we need to specify the initial conditions $S(0)$ (many), $I(0)$ (few), and $R(0)$ (=0), as well as the parameters β and ν.

In practical applications of the model, estimating the parameters can be a major challenge. We can estimate ν from the fact that $1/\nu$ is the average recovery time for the disease, which is usually possible to determine from early cases. However, estimating the infection rate β is more challenging, as it encompasses numerous biological and sociological factors into a single value. It depends on the infectiousness of the disease and the interactions within the population, and is usually challenging to estimate for a new disease. In a global pandemic, the behavior of the population varies among different countries and changes over time. Therefore, β often needs to be adapted to different regions and phases of the disease outbreak.

Epidemiologists often refer to the basic reproduction number $R0$ of an epidemic, which represents the average number of new individuals infected by a single infected person. The critical value is $R0 = 1$, since an epidemic will decline if $R0 < 1$, and it will grow exponentially if $R0 > 1$. In the simple model we are considering here, the relationship between $R0$ and β is given by $R0 = \beta/\nu$, since β measures the number of disease transmissions per time, and $1/\nu$ is the mean duration of the infectious period. Be aware of the potential confusion between the R category in the SIR model, in particular its initial value $R(0)$, and the basic reproduction number $R0$. These quantities, $R0$ and

---

[1] A simpler version of the SIR model is also commonly used, where the disease transmission term is not scaled with N. Eq. (5.8) then reads $S' = -\beta SI$, and (5.8) is modified similarly. Since N is constant the two models are equivalent, but the version in (5.7)-(5.9) is more common in real-world applications and gives a closer relation between β and common parameters such as the reproduction number.

β, are not directly related, and the notation may not be optimal. However, we use it here because it is widely established in the field of epidemiology.

Although the system (5.7)-(5.9) appears simple, it is not easy to derive analytical solutions. For specific applications, simplifications can often be made to allow for simple analytical solutions. For instance, when studying the early phase of an epidemic, the focus is usually on the I category Since the number of infected cases is low compared with the entire population during this phase, it is reasonable to assume that S is approximately constant and equal to N. By substituting $S \approx N$ into (5.8), we obtain a simple equation describing exponential growth with the solution

$$I(t) = I_0 e^{(\beta - \nu)}. \tag{5.10}$$

Such an approximate formula may be very useful, in particular for estimating the model's parameters. In the early phase of an epidemic, the number of infected people typically follows an exponential curve, and we can fit the model's parameters so that (5.10) fits the observed dynamics. We can also relate the behavior of this simple model to the basic reproduction number $R0$ introduced above. With $R0 = \beta/\nu$, $R0 > 1$ results in a positive exponent in (5.10), while $R0 < 1$ leads to a negative exponent and a decline in $I(t)$. However, to fully describe the dynamics of the epidemic, we need to solve the complete system of ODEs, which requires numerical solvers like the ones developed in the previous chapters.

Solving the SIR Model with the ODESystem Class Hierarchy. We can easily solve the SIR model given (5.7)-(5.9) using the solver tools developed in the previous chapters. For typical parameter values the models are not stiff, and the explicit RK solvers work well. For instance, a simple code which implements the SIR model as a function, and solves it using the fourth-order RK method, may look as follows:

```
from ODESolver import RungeKutta4
import numpy as np
import matplotlib.pyplot as plt

def SIR_model(t, u):
    beta = 0.001
    nu = 1 / 7.0
    S, I, R = u[0], u[1], u[2]
    dS = -beta * S * I
    dI = beta * S * I - nu * I
    dR = nu * I
    return [dS, dI, dR]

S0 = 1000
I0 = 1
R0 = 0

solver = RungeKutta4(SIR_model)
solver.set_initial_condition([S0, I0, R0])
```

```
t_span = (0, 100)
t, u = solver.solve(t_span, N=101)
S = u[:, 0]
I = u[:, 1]
R = u[:, 2]

plt.plot(t, S, t, I, t, R)
plt.show()
```

The resulting plot is shown in Figure 5.2.

A Class Implementation of the SIR Model. As noted above, estimating
the parameters in the model is often challenging. In fact, one of the most
important applications of models like these is to predict the dynamics of
new and unknown diseases, such as during the global Covid-19 pandemic.
Accurate predictions of the number of disease cases are crucial for planning
an effective response to the epidemic. However, for a new disease most of the
model parameters are unknown, and the lack of data makes them challenging
to estimate. There are ways to estimate the parameters from the early disease
dynamics, but the estimates will contain a large degree of uncertainty, and
a common strategy to gain insight into the disease dynamics is to run the
model for multiple parameter sets, to explore different outbreak scenarios. We
can easily run the code above for multiple values of `beta` and `nu`. However,
it is inconvenient that both parameters are hardcoded as local variables in
the `SIR_model` function, since this requires us to manually edit the code
for each new parameter value we want to use. As we have seen earlier, it
is much better to represent such a parameterized function as a class. In a
class, the parameters can be set in the constructor, and the function itself is
implemented in a `__call__` method. A class for the SIR model could look
like:

```
class SIR:
    def __init__(self, beta, nu):
        self.beta = beta
        self.nu = nu

    def __call__(self, t, u):
        S, I, R = u[0], u[1], u[2]
        dS = -self.beta * S * I
        dI = self.beta * S * I - self.nu * I
        dR = self.nu * I
        return [dS, dI, dR]
```

As with the models discussed in earlier chapters, the use of the class is very
similar to that of the `SIR_model` function above. We create an instance of
the class with specific values of `beta` and `nu`, and then this instance can be
passed to the ODE solver just like any regular Python function.

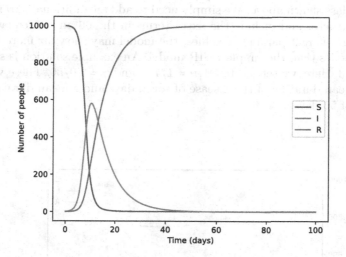

Fig. 5.2 Solution of the simplest version of the SIR model, showing how the number of people in each category (S, I, and R) changes with time.

5.2 Extending the SIR Model

The SIR model itself, in its simplest form, is rarely used for predictive simulations of real-world diseases. However, various extensions of the model are widely used to better capture the dynamics of different infectious diseases. In this section, we will explore a few such extensions that are based on the building blocks of the simple SIR model.

An SIR Model without Life-Long Immunity. One modification of the model is to remove the assumption of life-long immunity. The original model (5.7)-(5.9) describes a one-directional flow towards the R category, where the entire population eventually transitions to R if the model is solved over a sufficiently long time interval. However, this situation is not realistic for many diseases, since immunity tends to diminish over time. In the model this loss can be described by a leakage of people from the R category back to S. If we introduce the parameter γ to describe this flux ($1/\gamma$ being the mean time for immunity), the modified equation system looks like

$$S'(t) = -\beta SI/N + \gamma R,$$
$$I'(t) = \beta SI/N - \nu I,$$
$$R'(t) = \nu I - \gamma R.$$

Similar to the original model, the reduction in R is accompanied by an increase in S of the same magnitude, maintaining a constant total population

of $S + I + R$. The model can be implemented by a straightforward extension
of the SIR class shown above. We simply need to add an additional parameter
to the constructor and include the extra terms in the dS and dR equations.
By choosing different parameter values, the model may show far more interesting dynamics than the simplest SIR model. An example solution is shown
in Figure 5.3. Here, we set $\beta = 0.001, \nu = 1/7.0$, and $\gamma = 1.0/50$. These values
assume a mean duration of the disease of seven days and a mean duration of
immunity of 50 days.

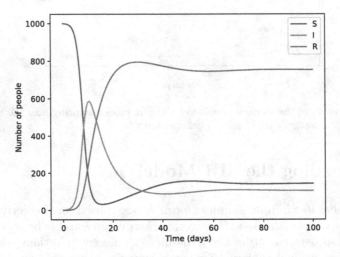

Fig. 5.3 Illustration of a SIR model without lifelong immunity, where people move
from the R category back to S after a given time.

A SEIR Model to Capture the Incubation Period. A SEIR model can
be used to account for the incubation period observed in many important
infections. During this period, individuals have been infected but are not
yet infectious themselves. To incorporate this aspect into the model, we can
add an additional category, E (for exposed). When people are infected, they
move into the E category rather than directly transitioning to the infected (I)
state. They then gradually move over to the infected state, where they can
also infect others. The model for how susceptible people get infected remains
the same as in the ordinary SIR model. Such a SEIR model is illustrated in
Figure 5.4, and the ODEs may look like

$$S'(t) = -\beta SI/N + \gamma R,$$
$$E'(t) = \beta SI/N - \mu E,$$
$$I'(t) = \mu E - \nu I,$$
$$R'(t) = \nu I - \gamma R.$$

Note that the overall structure of the model remains the same. Since the total population is conserved, all terms are balanced in the sense that they occur twice in the model, with opposite signs. A decrease in one category is always matched with an identical increase in another category. It is always useful to be aware of such fundamental properties in a model, since they can easily be checked in the computed solutions and may reveal errors in the implementation.

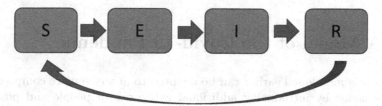

Fig. 5.4 Illustration of the SEIR model, without life-long immunity.

Again, this small extension of the model does not make it much more difficult to solve. The following code shows an example of how the SEIR model can be implemented as a class and solved with the `ODESolver` hierarchy:

```python
from ODESolver import RungeKutta4
import numpy as np
import matplotlib.pyplot as plt

class SEIR:
    def __init__(self, beta, mu, nu, gamma):
        self.beta = beta
        self.mu = mu
        self.nu = nu
        self.gamma = gamma

    def __call__(self, t, u):
        S, E, I, R = u
        N = S + I + R + E
        dS = -self.beta * S * I / N + self.gamma * R
        dE = self.beta * S * I / N - self.mu * E
        dI = self.mu * E - self.nu * I
        dR = self.nu * I - self.gamma * R
        return [dS, dE, dI, dR]

S0 = 1000
E0 = 0
```

```
I0 = 1
R0 = 0
model = SEIR(beta=1.0, mu=1.0 / 5, nu=1.0 / 7, gamma=1.0 / 50)

solver = RungeKutta4(model)
solver.set_initial_condition([S0, E0, I0, R0])
t_span = (0, 100)
t, u = solver.solve(t_span, N=101)
S = u[:, 0]
E = u[:, 1]
I = u[:, 2]
R = u[:, 3]

plt.plot(t, S, t, E, t, I, t, R)
plt.show()
```

5.3 A Model of the Covid-19 Pandemic

The models mentioned earlier can be adapted to describe more complex disease behavior by introducing additional categories of people and possibly more interactions between these categories. Now, we will explore an extension of the SEIR model, which was used by Norwegian health authorities to predict the spread of the Covid-19 pandemic throughout 2020 and 2021. In this case, we will derive the model as a system of ODEs, similar to the models discussed earlier, while the actual model used for providing Covid-19 predictions to health authorities was a stochastic model.[2] Stochastic models offer greater flexibility than the deterministic ODE version since they account for the inherent randomness and variability in disease transmission. In a stochastic SIR model, the disease transmission is considered a stochastic process, meaning that the probability of an individual getting infected is not fixed but depends on random events and chance encounters with infected individuals. Instead of using deterministic equations to model the number of individuals in each compartment and transitions between compartments, , stochastic models employ probability distributions and model transitions as stochastic processes rather than a continuous flux described by ODEs. One advantage of stochastic models is that they can more easily incorporate dynamics such as model parameters that vary with time after infection. For example, the infectiousness (β) typically follows a bell-shaped curve, gradually increasing after infection, reaching a peak after a few days, and then declining. Such behavior is easier to incorporate in a stochastic model compared with the deterministic ODE model considered here, which assumes a constant β for all individuals in the I category. However, the overall structure and dynamics of the two model types are exactly the same, and under

---

[2]See https://github.com/folkehelseinstituttet/spread

certain choices of model parameters, the stochastic and deterministic models become equivalent. For a discussion on stochastic and deterministic epidemiology models, refer to [6].

An important characteristic of Covid-19 is that people may be infected and capable of infecting others, even if they exhibit no symptoms. This attribute greatly impacts the spread of the disease since infected individuals are unaware of their condition and therefore do not take precautions to prevent transmission. There are two distinct groups of asymptomatic yet infectious people: have been identified:

- A certain number of people are infected, but never develop any symptoms, or the symptoms are so mild that they are mistaken for other mild respiratory infections. These asymptomatic people can still infect others, but with a lower infectiousness than the symptomatic group. Hence they need to be treated as a separate category.
- The other group, which is potentially even more significant for disease dynamics, consists of people who are infected and will develop symptoms, but the symptoms have not yet surfaced. However, they are still capable of infecting others, unlike the individuals in the *exposed* (E) category of the simple SEIR model above.

To model these two groups, we introduce two new compartments to the SEIR model presented earlier. We split the exposed category in two, E_1 and E_2, where the former represents non-infectious individuals and the latter represents individuals capable of infecting others. Similarly, we divide the I category into a symptomatic I and an asymptomatic I_a group. The transition from S to E_1 follows a similar pattern as in the SEIR model. However, from E_1, people can follow one of two possible trajectories: some move to E_2, then to I and finally to R, while others directly transition to I_a and then to R. The model is illustrated in Figure 5.5. Since there are two different E-categories and two different I-categories, we refer to the model as a SEEIIR model.

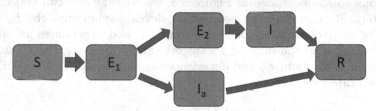

Fig. 5.5 Illustration of the Covid-19 epidemic model, with two alternative disease trajectories.

The derivation of the model equations for the SEEIR model is similar to the simpler models discussed earlier, but there are more equations and more terms involved. The most important extension in the SEEIR model is the inclusion of three categories of infectious people; $E_2, I,$ and I_a. Each of these categories interacts with the S category to create new infections, which we model in the same way as before. In a time interval Δt, the flux from S to E_1 consists of three contributions:

- Infected by people in I: $\beta \Delta t SI/N$
- Infected by people in I_a: $r_{ia}\beta \Delta t SI_a/N$
- Infected by people in E_2: $r_{e2}\beta \Delta t SE_2/N$

We allow for different infectiousness levels among the three categories, incorporated through a main infectiousness parameter β and two parameters r_{ia}, r_{e2} that scale the infectiousness for the two respective groups. By considering all three contributions and following the same steps as before to construct a difference equation and then an ODE, we obtain the following equation for the S category:

$$S'(t) = -\beta\frac{SI}{N} - r_{ia}\beta\frac{SI_a}{N} - r_{e2}\beta\frac{SE_2}{N}. \tag{5.11}$$

When people get infected they move from S to E_1. Therefore, the same three terms must appear in the equation for E_1, with opposite signs. Additionally, people in E_1 will move either to E_2 or I_a. Hence, we have

$$E_1'(t) = \beta\frac{SI}{N} + r_{ia}\beta\frac{SI_a}{N} + r_{e2}\beta\frac{SE_2}{N} - \lambda_1(1-p_a)E_1 - \lambda_1 p_a E_1$$
$$= \beta\frac{SI}{N} + r_{ia}\beta\frac{SI_a}{N} + r_{e2}\beta\frac{SE_2}{N} - \lambda_1 E_1.$$

Here, p_a is a parameter that describes the proportion of infected people who never develop symptoms, while $1/\lambda_1$ represents the mean duration of the non-infectious incubation period. The term $\lambda_1(1-p_a)E_1$ represents people moving to E_2, and $\lambda_1 p_a E_1$ are people moving to I_a. In the equation for E_1 we can combine these two fluxes into a single term, but they must be considered separately in the equations for E_2 and I_a.

Moving to the next step in Figure 5.5, we consider the two trajectories separately. Starting with the people that develop symptoms, the E_2 compartment receives an influx of people from E_1, and experiences an outflux as people move to the infected I category. Simultaneously, the I category receives an influx from E_2 and experiences an outflux to R. The ODEs for these two categories become

$$E_2'(t) = \lambda_1(1-p_a)E_1 - \lambda_2 E_2,$$
$$I'(t) = \lambda_2 E_2 - \mu I,$$

where $1/\lambda_2$ and $1/\mu$ represent the mean durations of the E_2 and I phases, respectively. The model for the asymptomatic disease trajectory is somewhat simpler, with I_a receiving an influx from E_1 and losing people directly to R. We have

$$I_a'(t) = \lambda_1 p_a E_1 - \mu I_a,$$

where we have assumed that the duration of the I_a period is the same as for I, i.e., $1/\mu$. Finally, the dynamics of the recovered category are governed by

$$R'(t) = \mu I + \mu I_a.$$

Note that we do not consider flow from the R category back to S, so we have effectively assumed life-long immunity. This assumption is not correct for Covid-19, but in the early phase of the pandemic, the duration of immunity was largely unknown, and the loss of immunity was therefore not considered in the models.

To summarize, the complete ODE system of the SEEIIR model can be written as

$$
\begin{aligned}
S'(t) &= -\beta\frac{SI}{N} - r_{ia}\beta\frac{SI_a}{N} - r_{e2}\beta\frac{SE_2}{N}, \\
E_1'(t) &= \beta\frac{SI}{N} + r_{ia}\beta\frac{SI_a}{N} + r_{e2}\beta\frac{SE_2}{N} - \lambda_1 E_1, \\
E_2'(t) &= \lambda_1(1-p_a)E_1 - \lambda_2 E_2, \\
I'(t) &= \lambda_2 E_2 - \mu I, \\
I_a'(t) &= \lambda_1 p_a E_1 - \mu I_a, \\
R'(t) &= \mu(I + I_a).
\end{aligned}
$$

A suitable choice of default parameters for the model can be as follows:

Parameter	Value
β	0.33
r_{ia}	0.1
r_{e2}	1.25
λ_1	0.33
λ_2	0.5
p_a	0.4
μ	0.2

These parameters are similar to the ones used by the health authorities to model the early phase of the Covid-19 outbreak in Norway. During this time, the behavior of the disease was largely unknown, and estimating the number of cases in the population was challenging. Consequently, fitting the parameter values was difficult, and they carried considerable uncertainty. As mentioned earlier, the most challenging parameters to estimate are those related to infectiousness and disease spread, which in this model are β, r_{ia}, and r_{e2}. Throughout the course of the pandemic, these parameters have been updated multiple times to reflect new knowledge about the disease and actual changes in disease spread due to new mutations or shifts in population behavior.

It is worth noting that we have set $r_{e2} > 1$, indicating that people in the E_2 category are more infectious than the infected group in I. This assumption reflects the fact that the E_2 group is asymptomatic, so people in this group are likely to be more mobile and potentially infect more people than those in the I group. On the other hand, the I_a group is also asymptomatic and therefore likely to have normal social interactions, but it is assumed that these people have a very low virus count. They are therefore less infectious than the people that develop symptoms, which is reflected in the low value of r_{ia}.

The parameters μ, λ_1, and λ_2 are given in units of days^{-1}, Thus the mean duration of the symptomatic disease period is five days ($1/\mu$), the non-infectious incubation period lasts three days on average ($1/\lambda_1$), while the mean duration of the infectious incubation period (E_2) is two days ($1/\lambda_2$). In this model, with multiple infectious categories, the basic reproduction number is calculated as

$$R0 = r_{e2}\beta/\lambda_2 + r_{ia}\beta/\mu + \beta/\mu,$$

since the mean duration of the E_2 period is $1/\lambda_2$ and the mean duration of both I and Ia is $1/\mu$. The parameter choices listed above yield $R0 \approx 2.62$, which is the value used by the Institute of Public Health (FHI) to model the early stage of the outbreak in Norway, from mid-February to mid-March 2020.

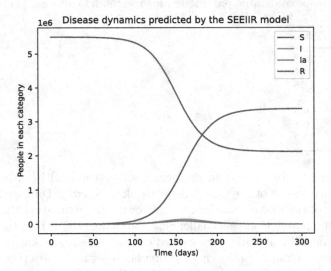

Fig. 5.6 Solution of the SEEIIR model with the default parameter values, which are similar to the values used by Norwegian health authorities during the early phase of the Covid-19 pandemic.

Although the present model is somewhat more complex than the previous ones, the implementation is not very different. A class implementation may look as follows:

```
class SEEIIR:
    def __init__(self, beta=0.33, r_ia=0.1,
                       r_e2=1.25, lmbda_1=0.33,
                       lmbda_2=0.5, p_a=0.4, mu=0.2):

        self.beta = beta
        self.r_ia = r_ia
        self.r_e2 = r_e2
        self.lmbda_1 = lmbda_1
        self.lmbda_2 = lmbda_2
        self.p_a = p_a
        self.mu = mu

    def __call__(self, t, u):
        beta = self.beta
        r_ia = self.r_ia
        r_e2 = self.r_e2
        lmbda_1 = self.lmbda_1
        lmbda_2 = self.lmbda_2
        p_a = self.p_a
        mu = self.mu

        S, E1, E2, I, Ia, R = u
        N = sum(u)
        dS = -beta * S * I / N - r_ia * beta * S * Ia / N \
             - r_e2 * beta * S * E2 / N
        dE1 = beta * S * I / N + r_ia * beta * S * Ia / N \
              + r_e2 * beta * S * E2 / N - lmbda_1 * E1
        dE2 = lmbda_1 * (1 - p_a) * E1 - lmbda_2 * E2
        dI = lmbda_2 * E2 - mu * I
        dIa = lmbda_1 * p_a * E1 - mu * Ia
        dR = mu * (I + Ia)
        return [dS, dE1, dE2, dI, dIa, dR]
```

The model can be solved with any of the methods available in the ODESolver hierarchy, similar to the simpler models discussed earlier. An example solution with the default parameter values is shown in Figure 5.6. It is important to note that since the parameters listed above are based on the initial stage of the pandemic when no restrictions were in place, this solution may be interpreted as a potential worst case scenario for the pandemic in Norway if no government-imposed restrictions were implemented.

While the plot for the I category may not appear too dramatic at first glance, a closer inspection reveals that the peak reaches slightly above 140,000 people. Considering the limited knowledge available at that stage, particularly regarding the severity of Covid-19, it is not surprising that a scenario of 140,000 people being infected simultaneously caused concern among health authorities. Another interesting observation from the curve is that the S cat-

egory flattens out well below the total population number. This behavior exemplifies the concept of herd immunity, wherein when a sufficient number of people are immune to the disease, it effectively stops spreading even if many people remain susceptible. As we are aware, severe restrictions were put in place in most countries during the early spring of 2020, making it impossible to determine whether this worst case scenario would ever have materialized. To accurately capture the actual dynamics of the pandemic in Norway, we would need to incorporate the effect of societal changes and altered infectiousness over time by making the β parameter a function of time. For instance, we could define it as a piecewise constant function to match the observed trends in the data.

Appendix A
Programming of Difference Equations

Although the main focus of these notes is on solvers for *differential equations*, we find it useful to include a chapter on the closely related class of problems known as *difference equations*. The main motivation for including this topic in a book on ODEs is to highlight the similarity between the two classes of problems, and in particular, the similarity of the solution methods and their implementation. Indeed, solving ODEs numerically can be seen as a two-step procedure. First, a numerical method is applied to turn *differential equations* into *difference equations*, and then these equations are solved using simple for-loop. The standard formulation of difference equations is very easy to translate into a computer program, and some readers may find it easier to study these equations first, before moving on to ODEs. In the present chapter we will also touch upon famous sequences and series, which have important applications both in the numerical solution of ODEs and elsewhere.

A.1 Sequences and Difference Equations

Sequences is a central topic in mathematics with important applications in numerical analysis and scientific computing. In the most general sense, a sequence is simply a collection of numbers:

$$x_0, \ x_1, \ x_2, \ ..., \ x_n, ...$$

For certain sequences, we can derive a formula that expresses the n-th number x_n as a function of n. For instance, consider the sequence of all odd numbers:

$$1, 3, 5, 7, ...$$

For this sequence, we can write a simple formula for the n-th term

$$x_n = 2n + 1,$$

J. Sundnes, *Solving Ordinary Differential Equations in Python*,
Simula SpringerBriefs on Computing 15,
https://doi.org/10.1007/978-3-031-46768-4

and we can use this formula to represent the complete sequence in a compact form;

$$(x_n)_{n=0}^{\infty}, \quad x_n = 2n+1.$$

Other examples of sequences include

$$1,\ 4,\ 9,\ 16,\ 25,\ \dots \quad (x_n)_{n=0}^{\infty},\ x_n = n^2,$$

$$1,\ \frac{1}{2},\ \frac{1}{3},\ \frac{1}{2},\ \dots \quad (x_n)_{n=0}^{\infty},\ x_n = \frac{1}{n+1},$$

$$1,\ 1,\ 2,\ 6,\ 24,\ \dots \quad (x_n)_{n=0}^{\infty},\ x_n = n!,$$

$$1,\ 1+x,\ 1+x+\frac{1}{2}x^2,\ 1+x+\frac{1}{2}x^2+\frac{1}{6}x^3,\ \dots \quad (x_n)_{n=0}^{\infty},\ x_n = \sum_{j=0}^{n} \frac{x^j}{j!}.$$

These examples are all formulated as infinite sequences, which is common in mathematics. However, in real-life applications, sequences are usually finite: $(x_n)_{n=0}^{N}$. Some familiar examples include the annual value of a loan or an investment.

In many cases, it is impossible to derive an explicit formula for the entire sequence, and x_n is instead defined by a relation involving x_{n-1} and possibly earlier terms. Such equations are called *difference equations*, and they can be challenging to solve with analytical methods, since computing the n-th term requires calculating the entire sequence $x_0, x_1, \dots, x_{n-1}$. Performing these compuations by hand or with a calculator can be tedious, but a computer can easily solve the equation using a for-loop. Combining sequences and difference equations with programming enables us to consider far more interesting and useful cases.

A Difference Equation for Computing Interest. Let us first consider a simple example of a difference equation for computing interest on an investment. In its simplest form, we put x_0 money in a bank at year 0, with an interest rate of p percent per year. What is the value after n years? If p is constant, the solution to this problem is given by the simple formula

$$x_n = x_0(1+p/100)^n,$$

so there is no need to formulate and solve the problem as a difference equation. However, simple generalizations, such as a non-constant interest rate, make this formula difficult to apply, while a formulation based on a difference equation remains applicable.

To formulate the problem as a difference equation, we observe that the amount x_{n+1} at year $n+1$ is the amount at year n plus the interest for year n. This gives the following relation between x_{n+1} and x_n:

$$x_{n+1} = x_n + \frac{p}{100}x_n.$$

To compute x_n, we can start with the known x_0, and compute $x_1, x_2, \ldots, x_n$. The procedure involves repeating a simple calculation many times, which is tedious to do by hand, but well suited for a computer. The complete program for solving this difference equation may look like:

```python
import numpy as np
import matplotlib.pyplot as plt
x0 = 100                        # initial amount
p = 5                           # interest rate
N = 4                           # number of years
index_set = range(N + 1)
x = np.zeros(len(index_set))

x[0] = x0
for n in index_set[1:]:
    x[n] = x[n - 1] + (p / 100.0) * x[n - 1]

plt.plot(index_set, x, 'ro')
plt.xlabel('years')
plt.ylabel('amount')
plt.show()
```

The three lines starting with x[0] = x0 form the core of the program. Here, we initialize the first element in our solution array with the known x0, and then enter the for-loop to compute the rest. The loop variable n runs from 1 to $N (= 4)$, and the formula inside the loop computes x[n] from the known x[n-1].

Also note that we pass a single array as an argument to plt.plot, whereas in most examples in this book we pass two arrays, typically representing time on the x-axis and the solution on the y-axis. When only one array of numbers is sent to plot, they are automatically interpreted as the y-coordinates of the points, and the x-coordinates will be the indices of the array, in this case the numbers from 0 to N.

Solving a Difference Equation without Using Arrays. The previous program stored the sequence as an array, which is convenient for programming the solver and allows us to plot the entire sequence. However, if we are only interested in the solution at a single point, i.e., x_n, there is no need to store the entire sequence. Since each x_n only depends on the previous value x_{n-1}, we only need to store the last two values in memory. A complete loop can look like this:

```python
x_old = x0
for n in index_set[1:]:
    x_new = x_old + (p / 100.) * x_old
    x_old = x_new  # x_new becomes x_old at next step
print('Final amount: ', x_new)
```

For this simple case we can actually make the code even shorter, since x_old is only used in a single line, and we can instead simply overwrite the old value of x once it has been used:

```
x = x0              #x is here a single number, not array
for n in index_set[1:]:
    x = x + (p / 100.) * x
print('Final amount: ', x)
```

We can observe that these codes store just one or two numbers, and for each iteration of the loop, we simply update and overwrite the values we no longer need. While this approach is straightforward and saves memory by not storing the complete array, programming with an array x[n] is usually safer, and we are often interested in plotting the entire sequence. Therefore, in the subsequent examples, we will mostly use arrays.

Extending the Solver for the Growth of Money. Suppose we want to change our interest rate model to one where interest is added every day instead of every year. The daily interest rate is $r = p/D$, where p is the annual interest rate and D is the number of days in a year. A common model in business applies $D = 360$, but n counts exact (all) days. The difference equation that relates the amount on one day to the previous day remains the same:

$$x_n = x_{n-1} + \frac{r}{100}x_{n-1},$$

except that the yearly interest rate has been replaced by the daily (r). If we want to determine the growth of money between two given dates, we also need to find the number of days between those dates. This calculation can be done manually, but Python offers a convenient module named datetime for this purpose. The following session illustrates how it can be used:

```
>>> import datetime
>>> date1 = datetime.date(2017, 9, 29)  # Sep 29, 2017
>>> date2 = datetime.date(2018, 8, 4)   # Aug 4, 2018
>>> diff = date2 - date1
>>> print(diff.days)
309
```

Putting these tools together, a complete program for daily interest rates may look like

```
import numpy as np
import matplotlib.pyplot as plt
import datetime

x0 = 100                              # initial amount
p = 5                                 # annual interest rate
r = p / 360.0                         # daily interest rate

date1 = datetime.date(2017, 9, 29)
date2 = datetime.date(2018, 8, 4)
diff = date2 - date1
N = diff.days
index_set = range(N + 1)
x = np.zeros(len(index_set))
```

```
x[0] = x0
for n in index_set[1:]:
    x[n] = x[n - 1] + (r / 100.0) * x[n - 1]

plt.plot(index_set, x)
plt.xlabel('days')
plt.ylabel('amount')
plt.show()
```

This program is slightly more sophisticated than the first one, but one may still argue that solving this problem with a difference equation is unnecessarily complex when we can simply apply the well-known formula $x_n = x_0(1+\frac{r}{100})^n$ to compute any x_n we want. However, we know that interest rates change quite often, and the formula is only valid for a constant r. On the other hand, for the program based on solving the difference equation, we only need minor modifications to handle a varying interest rate. The simplest approach is to let p be an array of the same length as the number of days, and fill it with the correct interest rates for each day. The modifications to the previous program may look like this:

```
p = np.zeros(len(index_set))
# fill p[n] with correct values

r = p / 360.0                          # daily interest rate
x = np.zeros(len(index_set))

x[0] = x0
for n in index_set[1:]:
    x[n] = x[n-1] + (r[n-1] / 100.0) * x[n-1]
```

The main difference from the previous example is that we initialize p as an array, and then r = p/360.0 becomes an array of the same length. In the formula inside the for-loop, we look up the correct value r[n-1] for each iteration of the loop. Filling p with the correct values can be non-trivial, but many cases can be handled quite easily. For instance, if the interest rate is piecewise constant and increases from 4.0% to 5.0% on a given date, the code for filling the array with values may look like this

```
date0 = datetime.date(2017, 9, 29)
date1 = datetime.date(2018, 2, 6)
date2 = datetime.date(2018, 8, 4)
Np = (date1 - date0).days
N = (date2 - date0).days

p = np.zeros(len(index_set))
p[:Np] = 4.0
p[Np:] = 5.0
```

A.2 More Examples of Difference Equations

As noted above, sequences, series, and difference equations have countless applications in mathematics, science, and engineering. Here we present a selection of well known examples.

Fibonacci Numbers as a Difference Equation. The sequence defined by the difference equation

$$x_n = x_{n-1} + x_{n-2}, \quad x_0 = 1, \; x_1 = 1,$$

is called the Fibonacci numbers. Originally derived for modeling rat populations, the Fibonacci numbers possess a range of interesting mathematical properties that have attracted considerable attention from mathematicians. The equation for the Fibonacci numbers differs from the previous examples, since x_n depends on the two previous values $(n-1, \, n-2)$, making it a second order difference equation. While this classification is important for mathematical solution techniques, the distinction between first and second order equations is minor in programming.

A complete code to solve the difference equation and generate the Fibonacci numbers can be written as

```
import sys
from numpy import zeros

N = int(sys.argv[1])
x = zeros(N+1, int)
x[0] = 1
x[1] = 1
for n in range(2, N+1):
    x[n] = x[n-1] + x[n-2]
    print(n, x[n])
```

In this code, we use the built-in list `sys.argv` from the `sys` model in order to provide the input N as a command-line argument. See, for instance, [16] for an explanation. It is important to note that we need to initialize both `x[0]` and `x[1]` before starting the loop, since the update formula involves both `x[n-1]` and `x[n-2]`. This is the main difference between this second order equation and the programs for first order equations considered above. The Fibonacci numbers grow quickly and running this program for large N will lead to overflow issues (try for instance $N = 100$). The NumPy `int` type supports up to 9223372036854775807, which is almost 10^{19}, so overflow is rarely a problem in practical applications. There are ways to avoid this issue, for instance using the standard Python `int` type instead of NumPy arrays, but we won't delve into those details here.

Logistic Growth. Returning to the initial problem of calculating the growth of money in a bank, we can write the classical solution formula more concisely as

$$x_n = x_0(1 + p/100)^n = x_0 C^n \quad (= x_0 e^{n \ln C}),$$

where $C = (1 + p/100)$. Since n represents years, this exemplifies exponential growth in time, following the general formula $x = x_0 e^{\lambda t}$. Similarly, populations of humans, animals, and other organisms exhibit the same type of growth when resources (such as space and food) are unlimited, and the exponential growth model has many applications in biology.[1] However, most environments can only support a finite number R of individuals, whereas the population continues to grow indefinitely in the exponential growth model. How can we modify the equation to create a more realistic model for growing populations?

Initially, when resources are abundant, we want the growth to be exponential, i.e., to grow with a given rate $r\%$ per year according to the difference equation:

$$x_n = x_{n-1} + (r/100)x_{n-1}.$$

To enforce the growth limit as $x_n \to R$, r must decay to zero as x_n approaches R. The simplest variation of $r(n)$ is linear:

$$r(n) = \varrho\left(1 - \frac{x_n}{R}\right)$$

We observe that $r(n) \approx \varrho$ for small n, when $x_n \ll R$, and $r(n) \to 0$ as n grows and $x_n \to R$. This formulation of the growth rate leads to the logistic growth model:

$$x_n = x_{n-1} + \frac{\varrho}{100}x_{n-1}\left(1 - \frac{x_{n-1}}{R}\right).$$

This is a *nonlinear* difference equation, while all the examples considered earlier were linear. The distinction between linear and nonlinear equations is crucial for the mathematical analysis of the equations, but it does not make much difference when solving the equation in a program. To modify the interest rate program mentioned above to describe logistic growth, we can simply replace the line

```
x[n] = x[n-1] + (p / 100.0) * x[n-1]
```

by

```
x[n] = x[n-1] + (rho / 100) * x[n-1] * (1 - x[n-1] / R)
```

A complete program may look like

```
import numpy as np
import matplotlib.pyplot as plt
x0 = 100                    # initial population
rho = 5                     # growth rate in %
R = 500                     # max population (carrying capacity)
```

---

[1] As discussed in Chapter 1, the formula $x = x_0 e^{\lambda t}$ is the solution of the differential equation $dx/dt = \lambda x$, which illustrates the close relation between difference equations and differential equations.

```
N = 200                           # number of years

index_set = range(N+1)
x = np.zeros(len(index_set))

x[0] = x0
for n in index_set[1:]:
    x[n] = x[n-1] + (rho / 100) * x[n-1] * (1 - x[n-1] / R)

plt.plot(index_set, x)
plt.xlabel('years')
plt.ylabel('amount')
plt.show()
```

Note that the logistic growth model is more commonly formulated as an ODE, as we discussed in Chapter 1 For certain choices of numerical method and discretization parameters, the program for solving the ODE is identical to the program for the difference equation discussed here.

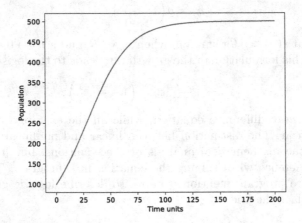

Fig. A.1 Solution of the logistic growth model for $x_0 = 100, \rho = 5.0, R = 500$.

The Factorial as a Difference Equation. The factorial $n!$ is defined as

$$n! = n(n-1)(n-2)\cdots 1, \quad 0! = 1 \qquad (A.1)$$

The following difference equation has $x_n = n!$ as solution and can be used to compute the factorial:

$$x_n = n x_{n-1}, \quad x_0 = 1$$

Similar to the interest rate example discussed earlier, one might question the usefulness of such a difference equation when we can simply use the formula (A.1) to compute the factorial for any value of n. However, in many applications, some of which will be discussed below, we need to compute the entire sequence of factorials $x_n = n!$ for $n = 0, \ldots N$. Although we could still apply (A.1) to compute each factor individually, it would involve many redundant computations, since we perform n multiplications for each new x_n. On the other hand, when solving the difference equation, each new x_n requires only a single multiplication, which can significantly speed up the program for large values of n.

Newton's Method as a Difference Equation. Newton's method is a popular method for solving nonlinear equations on the form

$$f(x) = 0.$$

Starting from some initial guess x_0, Newton's method gradually improves the approximation through iterations

$$x_n = x_{n-1} - \frac{f(x_{n-1})}{f'(x_{n-1})}.$$

We can recognize this as a nonlinear first-order difference equation. As $n \to \infty$, we hope that $x_n \to x_s$, where x_s is the solution to $f(x_s) = 0$. In practice, we solve the equation for $n \leq N$, for some finite N, just as for the difference equations considered earlier. But how do we choose N so that x_N is sufficiently close to the true solution x_s? Since we want to solve $f(x) = 0$, the best approach is to solve the equation until $f(x) \leq \epsilon$, where ϵ is a small tolerance. In practice, Newton's method usually converges rather quickly, or does not converge at all, so setting an upper bound on the number of iterations is a good idea. A simple implementation of Newton's method as a Python function may look like

```
def Newton(f, dfdx, x, epsilon=1.0E-7, max_n=100):
    n = 0
    while abs(f(x)) > epsilon and n <= max_n:
        x = x - f(x) / dfdx(x)
        n += 1
    return x, n, f(x)
```

The arguments f and dfdx are Python functions implementing $f(x)$ and its derivative. Both of these arguments are called inside the function and must therefore be callable. The x argument is the initial guess for the solution x, and the two optional arguments at the end are the tolerance and the maximum number of iterations. Although the method is implemented as a while-loop rather than a for-loop, the main structure of the algorithm remains the same as for the other difference equations considered earlier.

A.3 Systems of Difference Equations

So far, all the examples we have considered have been scalar difference equations, which describe how a single quantity changes from one step to the next. However, in many applications, it is necessary to track multiple variables simultaneously, and the dynamics of these variables may be coupled. This means that the value of one variable at step n depends on the values of multiple variables at step $n-1$. As an example, consider a simple extension of the interest rate model we discussed earlier. Assume that we have a fortune F invested with an annual interest rate of p percent, as before, but now we also want to consume an amount c_n every year. We can formulate a model to compute our fortune x_n at year n as a small extension of the previous difference equation. Simple reasoning tells us that the fortune at year n is equal to the fortune at year $n-1$ plus the interest minus the amount we spent in year $n-1$. Therefore, we have

$$x_n = x_{n-1} + \frac{p}{100}x_{n-1} - c_{n-1}.$$

In the simplest case, we can assume that c_n is constant, which would make this model a trivial extension of the interest rate model considered earlier. However, it is more natural to let c_n increase due to inflation. In this case, we obtain a system of difference equations describing the evolution of x_n and c_n. For instance, we may assume that c_n should grow with a rate of I percent per year, and in the first year we want to consume q percent of interest earned. The governing system of difference equations then becomes

$$x_n = x_{n-1} + \frac{p}{100}x_{n-1} - c_{n-1},$$
$$c_n = c_{n-1} + \frac{I}{100}c_{n-1}.$$

The initial conditions are $x_0 = F$ and $c_0 = (pF/100)(q/100) = \frac{pFq}{10000}$. This is a coupled system of two first-order difference equations, but the programming involved is not much more difficult than for the single equation we discussed earlier. We simply create two arrays x and c, initialize x[0] and c[0] with the given initial conditions, and then update x[n] and c[n] inside the loop. A complete code may look like this:

```
import numpy as np
import matplotlib.pyplot as plt
F = 1e7                          # initial amount
p = 5                            # interest rate
I = 3
q = 75
N = 40                                   # number of years
index_set = range(N + 1)
x = np.zeros(len(index_set))
```

```
c = np.zeros_like(x)

x[0] = F
c[0] = q * p * F * 1e-4

for n in index_set[1:]:
    x[n] = x[n - 1] + (p / 100.0) * x[n - 1] - c[n - 1]
    c[n] = c[n - 1] + (I / 100.0) * c[n - 1]

plt.plot(index_set, x, 'ro', label='Fortune')
plt.plot(index_set, c, 'go', label='Yearly consume')
plt.xlabel('years')
plt.ylabel('amounts')
plt.legend()
plt.show()
```

Another example of a system of difference equations is an extension of the logistic growth model we discussed earlier. While the logistic model describes the growth of a single population in the absence of predators, the famous Lotke-Volterra model describes the interaction of two species, a predator and a prey, in the same ecosystem. If we let x_n be the number of prey and y_n the number of predators on day n, the model for the population dynamics can be written as

$$x_n = x_{n-1} + ax_{n-1} - bx_{n-1}y_{n-1},$$
$$y_n = y_{n-1} + dbx_{n-1}y_{n-1} - cy_{n-1}.$$

Here, a is the natural growth rate of the prey in the absence of predators, b is the death rate of prey per encounter of prey and predator, c is the natural death rate of predators in the absence of food (prey), and d is the efficiency of turning predated prey into predators. This is a system of two first-order difference equations, similar to the previous example, and a complete solution code may look as follows.

```
import numpy as np
import matplotlib.pyplot as plt

x0 = 100                    # initial prey population
y0 = 8                      # initial predator pop.
a = 0.0015
b = 0.0003
c = 0.006
d = 0.5
N = 10000                   # number of time units (days)
index_set = range(N + 1)
x = np.zeros(len(index_set))
y = np.zeros_like(x)

x[0] = x0
y[0] = y0
```

```
for n in index_set[1:]:
    x[n] = x[n - 1] + a * x[n - 1] - b * x[n - 1] * y[n - 1]
    y[n] = y[n - 1] + d * b * x[n - 1] * y[n - 1] - c * y[n - 1]

plt.plot(index_set, x, label='Prey')
plt.plot(index_set, y, label='Predator')
plt.xlabel('Time')
plt.ylabel('Population')
plt.legend()
plt.show()
```

A.4 Taylor Series and Approximations

Sequences and series are extremely useful for approximating functions. For instance, commonly used functions like $\sin x, \ln x$, and e^x have been defined to have some desired mathematical properties, and we have an intuitive understanding of how they look, but we need an algorithm to evaluate the function values. One convenient approach is to approximate these functions using polynomials, since they are easy to calculate. Polynomial approximations have been used for centuries to compute exponentials, trigonometric functions and others. The most famous and widely used series for such approximations are the Taylor series, discovered in 1715, and given by

$$f(x) = \sum_{k=0}^{\infty} \frac{1}{k!} \left(\frac{d^k f(0)}{dx^k} \right) x^k. \tag{A.2}$$

Here, the notation $d^k f(0)/dx^k$ means the k-th derivative of f evaluated at $x = 0$. We can calculate a few of the terms in the sum to get

$$f(x) = f(0) + f'(0)x + \frac{1}{2}f''(0)x^2 + \frac{1}{6}f'''(0)x^3 \dots,$$

which makes it obvious that the right-hand side of (A.2) is in fact a polynomial in x. This means that for any function $f(x)$, if we can compute the function value and its derivatives for $x = 0$, we can approximate the function value at any x by evaluating a polynomial. In practice, we always work with a truncated version of the Taylor series:

$$f(x) \approx \sum_{k=0}^{N} \frac{1}{k!} \left(\frac{d^k f(0)}{dx^k} \right) x^k. \tag{A.3}$$

The accuracy of the approximation improves as N is increased. However, the most popular choice is $N = 1$, which provides a reasonable approximation close to $x = 0$ and has been essential in developing physics and technology.

We can also shift the variables to make these truncated Taylor series accurate around any value $x = a$:

$$f(x) \approx \sum_{k=0}^{N} \frac{1}{k!} \left(\frac{d^k f(a)}{dx^k} \right) (x-a)^k.$$

One of many applications of truncated Taylor series is to derive numerical methods for ODEs, and to analyze their accuracy, as we briefly introduced in Chapter 2.

As an example, let us consider the exponential function. Since we know that $d^k e^x / dx^k = e^x$ for all k, and $e^0 = 1$, we can substitute these values into (A.3) to get

$$e^x = \sum_{k=0}^{\infty} \frac{x^k}{k!}$$

$$\approx \sum_{k=0}^{N} \frac{x^k}{k!}.$$

Choosing, for instance, $N = 1$ and $N = 4$, we get the approximations

$$e^x \approx 1 + x,$$

$$e^x \approx 1 + x + \frac{1}{2} x^2 + \frac{1}{6} x^3.$$

These approximations are not very accurate for large x, but close to $x = 0$ they are sufficiently accurate for many applications. We can construct Taylor series approximations for other functions using similar arguments. For instance, consider $sin(x)$, where the derivatives follow the repetitive pattern $\sin'(x) = \cos(x), \sin''(x) = -\sin(x), \sin'''(x) = -\cos(x), \ldots \ldots$. We also have $\sin(0) = 0, \cos(0) = 1$. In general, we have $d^k \sin(0)/dx^k = (-1)^k mod(k,2)$, where $mod(k,2)$ is zero for k even and

$$\sin x = \sum_{k=0}^{\infty} (-1)^k \frac{x^{2k+1}}{(2k+1)!}.$$

Taylor Series Formulated as Difference Equations. We consider again the Taylor series for e^x around $x = 0$, given by

$$e^x = \sum_{k=0}^{\infty} \frac{x^k}{k!}.$$

If we define e_n as the approximation with n terms, i.e. for $k = 0, \ldots, n-1$, we can write

$$e_n = \sum_{k=0}^{n-1} \frac{x^k}{k!} = \sum_{k=0}^{n-2} \frac{x^k}{k!} + \frac{x^{n-1}}{(n-1)!},$$

and we can formulate the sum in e_n as the difference equation

$$e_n = e_{n-1} + \frac{x^{n-1}}{(n-1)!}, \quad e_0 = 0. \tag{A.4}$$

We see that this difference equation involves $(n-1)!$, which results in many redundant multiplications when computing the complete factorial for every iteration. However, we can use the idea of a difference equation for the factorial to compute the Taylor polynomial more efficiently. We have

$$\frac{x^n}{n!} = \frac{x^{n-1}}{(n-1)!} \cdot \frac{x}{n},$$

and if we let $a_n = x^n/n!$ it can be computed efficiently by solving

$$a_n = a_{n-1} \frac{x}{n}, \quad a_0 = 1.$$

Now we can formulate a system of two difference equations for the Taylor polynomial, where we update each term using the a_n equation and sum the terms via the e_n equation:

$$e_n = e_{n-1} + a_{n-1}, \quad e_0 = 0,$$
$$a_n = \frac{x}{n} a_{n-1}, \quad a_0 = 1.$$

Although we are solving a system of two difference equations, the computation is far more efficient than solving the single equation in (A.4) directly, since we avoid the repeated multiplications involved in the factorial computation.

A complete Python code for solving the system of difference equations and computing the approximation to the exponential function may look like

```
import numpy as np

x = 0.5  # approximate exp(x) for x = 0.5
N = 5
index_set = range(N + 1)
a = np.zeros(len(index_set))
e = np.zeros(len(index_set))
a[0] = 1

print(f'Exact: exp({x}) = {np.exp(x)}')
for n in index_set[1:]:
    e[n] = e[n - 1] + a[n - 1]
    a[n] = x / n * a[n - 1]
    print(f'n = {n}, appr. = {e[n]}, e = {np.abs(e[n]-np.exp(x)):4.5f}')
```

The output from this small program looks as follows:

```
Exact: exp(0.5) = 1.64872
n = 1, appr. = 1.00000, e = 0.64872
n = 2, appr. = 1.50000, e = 0.14872
n = 3, appr. = 1.62500, e = 0.02372
n = 4, appr. = 1.64583, e = 0.00289
n = 5, appr. = 1.64844, e = 0.00028
```

This program first prints the exact value e^x for $x = 0.5$, and then the Taylor approximation and associated error for $n = 1$ to $n = 5$. The Taylor series approximation is most accurate close to $x = 0$. Choosing a larger value of x would therefore lead to larger errors, and we would need to also increase n for the approximation to be accurate.

References

[1] U. M. Ascher and L. R. Petzold. *Computer Methods for Ordinary Differential Equations and Differential-Algebraic Equtions.* SIAM, 1998.

[2] C. F. Curtiss and J. O. Hirschfelder. Integration of stiff equations. *Proc. Nat. Acad. Sci.*, 38:235–243, 1952.

[3] Peter Deuflhard and Folkmar Bornemann. *Scientific Computing With Ordinary Differential Equations*, volume 42. Springer, 2012.

[4] J. R. Dormand and P. J. Prince. A family of embedded runge-kutta formulae. *J. Comput. Appl. Math.*, 6:19–26, 1980.

[5] E. Fehlberg. Klassische runge-kutta-formeln vierter und niedrigerer ordnung mit schrittweiten-kontrolle und ihre anwendung auf wärmeleitungsprobleme. *Computing*, 6(1):61–71, 1970.

[6] Priscilla E. Greenwood and Luis F. Gordillo. Stochastic epidemic modeling. In Gerardo Chowell, James M. Hyman, Luís M. A. Bettencourt, and Carlos Castillo-Chavez, editors, *Mathematical and Statistical Estimation Approaches in Epidemiology*, pages 31–52. Springer, 2009.

[7] Kjell Gustafsson. Using control theory to improve stepsize selection in numerical integration of ODE. *IFAC Proceedings Volumes*, 23(8):405–410, 1990.

[8] E. Hairer, S. P. Nørsett, and G. Wanner. *Solving Ordinary Differential Equations I, Nonstiff Problems.* Springer, 1991.

[9] E. Hairer and G. Wanner. *Solving Ordinary Differential Equations II, Stiff and Differential Algebraic Problems.* Springer, 1991.

[10] A.L. Hodgkin and A. F. Huxley. A quantitative description of of membrane current and its aplication to conduction and excitation in nerve. *J Physiol*, 117:500–544, 1952.

[11] J. Keener and J. Sneyd. *Mathematical Physiology.* Springer, 2009.

[12] WO Kermack and AG McKendrick. Contributions to the mathematical theory of epidemics-i. 1927. *Bulletin of mathematical biology*, 53(1-2), 1991.

[13] Anne Kværnø. Singly diagonally implicit runge-kutta methods with an explicit first stage. *BIT Numerical Mathematics*, 44:489–502, 2004.

© The Author(s) 2024
J. Sundnes, *Solving Ordinary Differential Equations in Python*,
Simula SpringerBriefs on Computing 15,
https://doi.org/10.1007/978-3-031-46768-4

[14] Hans Petter Langtangen and Hans Petter Langtangen. *A Primer on Scientific Programming With Python*, volume 6. Springer, 2012.

[15] S. Rush and H. Larsen. A practical algorithm for solving dynamic membrane equations. *IEEE Transactions on Biomedical Engineering*, 25(4):389–392, 1978.

[16] Joakim Sundnes. *Introduction to Scientific Programming With Python*. Springer, 2020.

Index

© The Author(s) 2024
J. Sundnes, *Solving Ordinary Differential Equations in Python*,
Simula SpringerBriefs on Computing 15,
https://doi.org/10.1007/978-3-031-46768-4

Printed in the United States
by Baker & Taylor Publisher Services